# WILDERNESS AND THE COMMON GOOD

## A NEW ETHIC OF CITIZENSHIP

JO ARNEY

Library of Congress Cataloging-in-Publication Data

Arney, Jo.
Wilderness and the common good : a new ethic of citizenship / Jo Arney.
pages cm
ISBN 978-1-936218-19-6
1. Political ecology. 2. Citizenship 3. Environmental policy--Citizen participation. 4. Common good. 5. Environmental responsibility. I. Title.
JA75.8.A76 2015
304.2--dc23

2015009130

Printed in the United States of America
0 9 8 7 6 5 4 3 2 1

Fulcrum Publishing
4690 Table Mountain Dr., Ste. 100
Golden, CO 80403
800-992-2908 • 303-277-1623
www.fulcrumbooks.com

## Praise for *Wilderness and the Common Good*

"Half a century after Congress passed the Wilderness Act, Jo Arney here reflects on how much our shared common good values wildlands. We dwell on common ground. Citizens can and ought to become residents of their lands. Three dimensional persons need landscapes urban, rural – and wild. Without the last, originally the first environment, we cannot value life to the fullest. Moving from civic policy to personal experience, her argument and appeal is at once penetrating and quite down to earth."

— **Holmes Rolston, III**,
University Distinguished Professor
and Professor of Philosophy,
Colorado State University

"Weaving together her personal story, her teaching experiences, and insightful political analysis, Jo Arney has created a thoughtful study of the relationship between citizenship and wilderness. Throughout her study, she makes an eloquent plea for a quest for common ground, rejecting the bitter divisiveness and polarization that characterizes so much of our national political life today. On the 50th anniversary of the creation of the Wilderness Act, Wilderness and the Common Good is a fitting and illuminating tribute."

— **George L. Mehaffy**,
Vice President,
American Association of
State Colleges and Universities (AASCU)

*In loving memory of Tom McCoy*

*During the years in which I got to know Tom, he was very committed to his work as the Wirth Chair in Sustainable Development at the University of Colorado–Denver. Tom introduced me to many of the people who have supported this project, and without him this book likely would never have happened. Not only was Tom a good friend to all who knew him, he was a generous spirit who saw the potential in others before they saw it in themselves. This most certainly includes me. Tom, you are dearly missed.*

# Contents

# ACKNOWLEDGMENTS

When I was invited to write this book, I was told that it was in part because I am an optimist. I admit to being guilty as charged. I do believe that the state of our wilderness is strong. I believe that young adults will be good stewards of our public lands. I believe these things because I have been inspired by so many others, many of them students. The young adults I have the privilege of teaching at the University of Wisconsin–La Crosse inspire me to believe in the power of younger generations to create a new and better world. One of the most meaningful moments of my career was the first time I took a group of students to study at Yellowstone and saw it through their eyes. Thank you to Elena Bantle, Tyler Burkart, Jacob Hanifl, Julie (Kildahl) Matucheski, Katelyn Larsen, Lynn Lodahl, Kevin McCuster, and Jessica (Verbist) Cooper for sharing this opportunity with me.

This book was also influenced by my own environmental philosophy instructors, including Michael Nelson at the University of Wisconsin–Stevens Point (currently at Oregon State) and Holmes Rolston at Colorado State University. I extend a special thank you to George Mehaffy, who created the Stewardship of Public Lands initiative of the American Association of State Colleges and Universities' American Democracy Project and who has generously allowed me to remain affiliated with it. Thank you also to Brad Bulin with the Yellowstone Association for sharing your love of Yellowstone with your students, including me, and thereby inspiring us to become stewards of the land. I also thank my colleagues in the Political Science and Public Administration Department at UW–L for creating a work environment in which we all can thrive. You are more than just colleagues, you are dear friends. Thank you especially to the members of our research group, C4: Jeremy Arney, Ray Block, Tim Dale, Christina Haynes, Steve McDougal, and Adam Van Liere. Last but not least, I want to thank my friends Sam Scinta and Kristen Foehner, who suggested I write this book. Thanks for believing that I had something to say that was worth writing.

Jo Arney

2015

# Introduction

*"There is a love of wild nature in everybody, an ancient mother-love ever showing itself whether recognized or no, and however covered by cares and duties."*
– John Muir

The Wilderness Act was signed into law by President Lyndon B. Johnson on September 3, 1964, after eight years of debate in the halls of Congress. Its architect, Howard Zahniser of The Wilderness Society, penned one of the most successful environmental laws in our nation's history and validated the work started by innumerable wilderness advocates before him. The fiftieth anniversary of the Wilderness Act, in 2014, witnessed more than 750 designated wilderness areas comprising over 109 million acres. These wild areas are not only important for the biodiversity that they safeguard, they are essential for the common good of our society.

In commemoration of the fortieth anniversary of the Wilderness Act, Doug Scott wrote *The Enduring Wilderness*. In it Scott provides a vibrant account of the background and creation of the Wilderness Act, including a detailed history of the forty-year battle to establish "wilderness" as a legal designation. He weaves in the stories of those who worked so hard to conserve our wildlands, from Aldo Leopold to Howard Zahniser, and highlights the importance of grassroots efforts in the crusade to protect wilderness.

In this book I take a slightly different approach to celebrating wilderness, one that explores intersecting themes found in advocacy, education, and citizenship. I draw on the work of environmental scientists, advocates, and philosophers to describe the role citizens can and should play in wilderness protection. I also include personal narratives that trace my own journey from growing up in rural Wisconsin to studying environmental philosophy and policy in Colorado to becoming an instructor of environmental politics at the University of Wisconsin–La Crosse. The experiences I describe in these vignettes have shaped both my love and appreciation of all things wild.

At its heart this book is about wilderness but also about education and citizenship. Protecting wilderness will take coalitions of citizens working together toward a common good. Being a citizen includes being educated about the world and active in one's community. Being a citizen also sometimes requires compromise: considering others' opinions, respecting their values, and looking for common ground. True protection for wilderness areas, both future and present, will be born out of a shared understanding about what saving it means for ourselves and our society.

I recently met with individuals from across the country who came together to define civic engagement. In many ways the word *citizen* is symbiotic with the concept of civic engagement. The discussion became heated, however, as the group debated whether or not to use the word *citizen* in the definition of civic engagement. Given the current controversies surrounding immigration in the United States, it is important to explore the concept of citizenship and how citizens relate to the common good. In this book I adopt this enlightened definition rather than the legalistic one. Here the word *citizen* does not refer to an individual born or naturalized in the United States, but rather to someone who cares about his or her community and the common good. Citizenship is a reflection of actions individuals take and beliefs they hold. A citizen is a person who promotes and sustains relationships among members of a community and helps to build social capital.

The challenge for wilderness advocates today lies not only in convincing Congress that certain places should be designated as wilderness, but also in engaging in a dialogue about why wilderness areas should be preserved and what preserving them means for our shared future.

## Chapter One
# The Wilderness Act and Its Legacy

*"In all the category of outdoor vocations and outdoor sports there is not one, save only the tilling of the soil, that bends and molds the human character like wilderness travel."*

– Aldo Leopold
"The Last Stand of Wilderness"

Being a steward of the land is an act of citizenship. A devoted citizen works to foster and protect the collective common good, which includes safeguarding our wilderness. There are hundreds of books and articles that seek to define and delineate wilderness, and in this book I refer often to the political definition. But it is important to understand that wilderness is many things to many people. It is a place humans visit but rarely dwell. It is a place to be admired, sometimes feared, often held sacred. Wilderness has many important roles in our lives, from serving as a reservoir for clean air and water to fueling renewal of the human spirit.

The political definition of wilderness can be found in the Wilderness Act of 1964:

> A wilderness, in contrast with those areas where man and his own works dominate the landscape, is hereby recognized as an area where the earth

> and its community of life are untrammeled by man, where man himself is a visitor who does not remain. An area of wilderness is further defined to mean in this Act an area of undeveloped Federal land retaining its primeval character and influence, without permanent improvements or human habitation, which is protected and managed so as to preserve its natural conditions and which (1) generally appears to have been affected primarily by the forces of nature, with the imprint of man's work substantially unnoticeable; (2) has outstanding opportunities for solitude or a primitive and unconfined type of recreation; (3) has at least five thousand acres of land or is of sufficient size as to make practicable its preservation and use in an unimpaired condition; and (4) may also contain ecological, geological, or other features of scientific, educational, scenic, or historical value.

Passing the Wilderness Act took a great deal of political will. The decision to designate an area as wilderness is made not by those who work alongside wild areas in our national parks and forests, but by elected officials in Washington, DC. According to the law, only Congress can designate a wilderness area, and only Congress can remove the designation or adjust its boundaries. This ensures that these national treasures are true public goods protected by representatives of the people.

In the last decade alone, Congress has designated 80 new wilderness areas by way of 13 new public laws. In the United States today there are 757 wilderness areas covering 109,511,966 acres of land. Approximately 44 percent of that land is managed by the National Park Service (NPS), 33 percent by the United States Forest Service (USFS), 19 percent by the United States Fish and Wildlife Service (USFWS), and 8 percent by the Bureau of Land Management (BLM).

## An Act of Congress

The authors of the Wilderness Act wanted to ensure that these designated areas were truly protected and thus required an act of Congress to designate an area as wilderness. If designation was left in the hands of park and forest administrators, they feared, it would be far too easy for special interests to influence the shifting of an area's boundaries or the removal of the designation altogether. The act took eight years to pass, in part because both Forest Service and National Parks administrators opposed the bill, fearing that it would obstruct their own discretion within national forest and park boundaries. The key compromise made to get the bill passed was the power given to Congress to remove the wilderness designation. While this provision was initially included to support those who opposed the act, in practice it has helped protect wilderness areas.

What made the act especially controversial was the proposed limitation of human activities after an area is designated wilderness. Only those activities that have a truly minimal impact, such as hiking or fly-fishing, were allowed. The act reads:

> For this purpose there is hereby established a National Wilderness Preservation System to be

> composed of federally owned areas designated by Congress as "wilderness areas," and these shall be administered for the use and enjoyment of the American people in such manner as will leave them unimpaired for future use as wilderness, and so as to provide for the protection of these areas, the preservation of their wilderness character, and for the gathering and dissemination of information regarding their use and enjoyment as wilderness. . . . (Sec 2(a))

> Except as specifically provided for in this Act, and subject to existing private rights, there shall be no commercial enterprise and no permanent road within any wilderness area designated by this Act and, except as necessary to meet minimum requirements for the administration of the area for the purpose of this Act (including measures required in emergencies involving the health and safety of persons within the area), there shall be no temporary road, no use of motor vehicles, motorized equipment or motorboats, no landing of aircraft, no other form of mechanical transport, and no structure or installation within any such area. (Sec 4(c))

## Opposition

Opponents of the Wilderness Act included those hoping to transform wild goods into something more valuable for human use. People who preferred a multitude of sustainable activities in wildlands also spoke out against the law. Many lawmakers argued that the value of wilderness is related to what citizens can make out of it. After all, even Aldo Leopold once noted that wilderness provided the raw materials out of which humans made civilization. Most frequently these objectors were mining and logging interests and representatives from states dependent on these industries. But there were other opponents as well, including many who wanted to maintain access to the wilderness for recreation. In response to revisions to the act in 1984, even some mountain biking organizations came out against the wilderness designation since limits on "mechanical transport" could be read as a provision against bicycles. Some opponents of the wilderness designation argued that the definition itself was confusing, while others contended that what counts as "substantially unnoticeable" work by humans is unclear. Other criticisms included that this conception of wilderness was too idealized, too ethnocentric, and devoid of the notion that wilderness naturally changes over time.

### Wilderness as an Overlay

The designation of wilderness is what's known as an overlay concept. Most of the lands currently designated as wilderness were already part of national parks or national forests when they were designated by Congress. As the act explains,

> The inclusion of an area in the National Wilderness Preservation System notwithstanding, the area shall continue to be managed by the

> Department and agency having jurisdiction thereover immediately before its inclusion in the National Wilderness Preservation System unless otherwise provided by Act of Congress. No appropriation shall be available for the payment of expenses or salaries for the administration of the National Wilderness Preservation System as a separate unit nor shall any appropriations be available for additional personnel stated as being required solely for the purpose of managing or administering areas solely because they are included within the National Wilderness Preservation System.

In fact, today 77 percent of all designated wilderness areas are managed by the National Park Service or the US Forest Service. When the overlay of wilderness is applied to an existing park, forest, or other plot of land, it applies only to the wild areas within those designated boundaries. In other words, even though parts of Yellowstone National Park contain designated wilderness areas, not all of Yellowstone is a wilderness area. It would be impossible for the whole park to be designated wilderness, since like most national parks today, Yellowstone has roads, lodges, concessions, and other developments not allowed under the Wilderness Act. The wilderness areas within national parks are often referred to as backcountry, and visitors are required to obtain a permit to travel within wilderness boundaries. About half of all US national park land is designated wilderness.

## Public Opinion Today

To mark the fortieth anniversary of the Wilderness Act, wilderness advocates publicly celebrated the sustained and growing public support of wilderness protection. A decade later, I have sought to capture the effects of that momentum by looking at public opinion polls over the past ten years. Unfortunately, few recent public opinion polls have included questions about wilderness. The iPOLL databank maintained by the Roper Center Public Opinion Archives (which tracks every major United States survey firm and more than 150 organizations) lists only seven questions in which *wilderness* was a key word since 2004. Perhaps not surprisingly, six out of those seven questions were asked in 2008 and focused on drilling in the Arctic National Wildlife Refuge (ANWR) or on drilling in wilderness areas in general. Drilling for oil in the ANWR was a hotly contested issue during the 2008 presidential election campaign, complete with chants of "Drill, baby, drill!" at the Republican National Convention. The results of the polls, however, are a mixed bag. When interpreting the results below, be certain to note the source of each poll.

### Gallup Poll, May 2008

(Please say whether you would favor or oppose taking each of the following steps to attempt to reduce the price of gasoline.) How about... allowing oil drilling in US (United States) coastal and wilderness areas now off-limits to oil exploration?

| | |
|---|---|
| **Favor** | **57%** |
| **Oppose** | **41%** |
| **No opinion** | **2%** |

## Democracy Corps Poll, June 2008

Given the high price of gas, would you favor or oppose allowing oil drilling in US (United States) coastal areas and wilderness areas that are currently protected by the federal government? (If Favor/Oppose, ask: Is that strongly or somewhat favor/oppose?)

| | |
|---|---|
| **Strongly Favor** | **38%** |
| **Somewhat Favor** | **17%** |
| **Somewhat Oppose** | **13%** |
| **Strongly Oppose** | **25%** |
| **Don't Know/Refuse** | **6%** |

## Unified Arctic Campaign Survey, June 2008
(2 questions)

We should not allow drilling in the Arctic National Wildlife Refuge because this is one of the most valuable wilderness areas left in the US (United States) and it would be permanently damaged by drilling... Agree, disagree.

| | |
|---|---|
| **Agree** | **56%** |
| **Disagree** | **38%** |
| **Don't Know/Refuse** | **6%** |

The oil companies already have obtained the right to drill for oil and gas in 32 million acres of oil-rich federal land that they are not using. As long as they have not drilled in the acres already available to them, there is no reason to open up more Alaskan wilderness to drilling... Agree, disagree.

| | |
|---|---|
| **Agree** | **68%** |
| **Disagree** | **24%** |
| **Don't Know/Refuse** | **8%** |

**ABC News/Planet Green/Stanford Poll, July 2008**

Do you think the federal government should or should not allow drilling for oil in US (United States) wilderness areas where it's currently not allowed?

| | |
|---|---|
| **Should** | **55%** |
| **Should not** | **43%** |
| **No opinion** | **2%** |

**CNN/Opinion Research Corporation Poll, July 2008**

(Please tell me whether you think each of the following is a major cause of the recent increase in gasoline prices, a minor cause, or not a cause at all?)... Federal laws that prohibit increased drilling for oil offshore or in wilderness areas:

| | |
|---|---|
| **Major** | **51%** |
| **Minor** | **32%** |
| **Not a cause** | **17%** |

The only recent poll question in the iPOLL database related to wilderness but not related to drilling was part of a March 2009 poll by the Climate Change and Global Poverty Survey: "What do you think is the most important environmental problem that the current (Obama) administration should focus on: climate change, air and water pollution, shrinking of wilderness areas or endangered species?" Clearly the authors of the question were not speaking solely about designated wilderness area but rather about wildlands in general. The results:

| | |
|---|---|
| **Climate change** | **23%** |
| **Air and water pollution** | **40%** |
| **Shrinking of wilderness areas** | **9%** |
| **Endangered species** | **4%** |
| **All of these** | **16%** |
| **Don't Know/Refused** | **8%** |

This last question may suffer from respondents having received a fixed set of answers to choose from. Nevertheless, wilderness protection was evidently not a priority among these participants. Taken together, what these poll questions (excluding the Unified Arctic Campaign Survey) demonstrate is that respondents in 2008 were more concerned about energy than they were about protecting wilderness.

## The Relationship between Wilderness and Mining

Americans are enormous consumers of energy. Setting aside the current green energy movement, energy production in this country has traditionally required some sort of resource extraction. In other words, the debate between energy production and wilderness protection is not new. And it is certainly not over. In June 2013 Republican governor Sean Parnell of Alaska publicly called for reopening the debate about drilling for oil and gas in the Arctic National Wildlife Refuge while addressing a national audience during a Sunday morning talk show. An astute reader might question why drilling is even open to debate in the ANWR and other wilderness areas. Surely mining would disturb wilderness areas more than mountain biking, for example. It is important to remember that passing the act required compromise, and one of the compromises made was for mineral rights. The act states:

> Notwithstanding any other provisions of this Act, until midnight December 31, 1983, the United States mining laws and all laws pertaining to mineral leasing shall, to the extent as applicable prior to September 3, 1964, extend to those national forest lands designated by this Act as "wilderness areas"; subject, however, to such reasonable regulations governing ingress and egress as may be prescribed by the Secretary of Agriculture consistent with the use of the land for mineral location and development and exploration, drilling, and production, and use of land for transmission lines, waterlines, telephone lines, or facilities necessary in exploring, drilling, producing, mining, and processing operations, including where essential the use of mechanized ground or air equipment and restoration as near as practicable of the surface of the land disturbed in performing prospecting, location, and, in oil and gas leasing, discovery work, exploration, drilling, and production, as soon as they have served their purpose.

Given the concessions underlying the Wilderness Act, debates about drilling in wilderness areas are likely to continue in public and in the halls of Congress. When Congress enacted the law it did not prohibit these activities from occurring, especially in areas where oil, gas, and mineral deposits were identified before wilderness designation.

## Lessons

There are lessons to be gleaned from the public opinion polls, in particular having to do with the shared ownership of these public lands. With shared ownership, of course, comes a variety of interests among the owners. When decisions made in Congress do not have a direct impact on an individual constituent, it is difficult to rally support from that individual. Moreover, the current size of our designated wilderness areas—over 100 million acres—is a concept almost incomprehensible to most people. When the public is told that a certain mining operation will alter only a few hundred acres, the land affected can easily be perceived as only a small fraction of designated wilderness and thus inconsequential. Finding common ground requires compromise, and a majority of Americans may view drilling as an acceptable compromise in order to meet our energy needs.

A second caveat involves the phenomenon of Not in My Backyard (NIMBY) when it comes to wilderness protection. Creating wilderness areas requires grassroots citizen activism, and citizens most want to protect the wildlands local to them—the lands they love. This may be an effective method for delineating wilderness, but it is also subject to an intensity problem. Chances are good that an individual who wants to protect one wilderness area will also want to protect others, but it is likely that he or she feels more strongly about the ones closer to home. Here's where the phenomenon of NIMBY comes into play. If we have to sacrifice some wilderness (in the interest of national energy, for example), maybe it is okay as long as it's happening somewhere else. Alaska is likely a very long way from the majority of the individuals answering poll questions asked by the mainstream US media. Individual opinions are also often framed by those in the political limelight. In earlier ANWR debates, Senator Ted Stevens from Alaska argued that the ANWR is nothing but a frozen wasteland for most of the year; Senator John Kerry countered that it is pristine wilderness. Depending on a person's opinions of the arguments put forth by Stevens or Kerry, he or she might not value the ANWR as highly as wilderness close to home. Given that 52 percent of all of our designated wilderness areas are in Alaska, it's not surprising that the majority of poll respondents favor drilling there. This intensity problem is why education and exposure are critically important in the fight for wilderness. We take care of what we value. It is perhaps even more telling that in the past decade there have been no public opinion poll questions about wilderness itself. If people aren't being asked about whether they value wilderness, then it is unlikely that those outside of the wilderness coalition are thinking much about it.

In its fifty-year history, the Wilderness Act has accomplished a great deal. It is one of the most successful and important pieces of legislation in the history of the United States. Unlike the majority of the legislation that followed during the pro-environment decade of the 1970s, the Wilderness Act was forward thinking and preventative. By contrast, the Clean Air and Clean

Water Acts sought to clean up what had already been polluted, as did the Comprehensive Environmental Response, Compensation, and Liability Act (also known as Superfund). The Endangered Species Act sought to restore species that were already in danger. But the Wilderness Act sought to protect the wild from possible dangers. It alone recognized the intrinsic value of our wildlands and created a mechanism for preserving them for the future.

Continuing this legacy of protection will require a lot of work for wilderness advocates yet to come. It will require dedicated citizens to act as stewards of the land, and in our ever more divisive and divided society, finding common ground will be vital to moving forward.

## Vignette
# Moving to the Farm

Skunk Hallow, as we affectionately call our little piece of property, lies just outside of Coon Valley, Wisconsin. This five-acre piece of land was once part of a working dairy and tobacco farm. It is located on a dead-end road and surrounded by the bluffs that make up southern Wisconsin's Driftless Area, so called because no glaciers reached the area during the last glacial period, leaving the land here characterized by bluffs and deeply carved river valleys, locally known as coulees. The word *coulee* is derived from Canadian French and literally means "to flow." It is no surprise, therefore, that Coon Valley became the site of the nation's first watershed conservation project in 1933. Aldo Leopold advised the Civilian Conservation Corps on the project and later wrote about it in *A Sand County Almanac*.

This is a mostly rural area sprinkled with farms and small communities that are dependent on agriculture. While admiring the beauty of the drive our first time out to view the property, I worried that it would take us too far from work. Having left Denver's traffic nightmare after almost nine years and moved to the small city of La Crosse, my husband and I had become accustomed to the ten-block walk to the university where we both work. We were unsure whether we were willing to endure a twenty-five-minute commute each way, especially given the snowy winters and the steep drive up and down the intermediate bluff. But as we looked at the property, something stirred in both of us. What we felt was more than a response to the sheer beauty of the location. And it was more than the enticement of returning to our roots, though both of us grew up in rural areas. What we felt was a connection with the land. An image emerged of not only what we could do there but who we would become by living and working on such a piece of property. We felt a sense of peace, of purpose, and of connection to mother earth.

I was certainly influenced by Aldo Leopold as I was growing up. I first read the beginning of *A Sand County Almanac* in high school and later memorized his land ethic statement while studying philosophy at the University of Wisconsin–Stevens Point. As Leopold so eloquently wrote, "A thing is right when it tends to preserve the integrity, stability, and beauty of the biotic community. It is wrong when it tends otherwise." My days at the farm remind me so much of the year Leopold chronicled in the first section of the book. Since moving here three years ago, I have had the occasion to watch deer, turkeys, raccoons, opossums, otters, minks, skunks, woodchucks, eagles, hawks, rabbits, toads, frogs, turtles, snakes, and countless other birds, animals, insects, and fish. Most evenings we can hear coyotes announcing their waking and their return to the den at dawn. The winter is full of footprints to explore.

There is no doubt Skunk Hallow is in a coulee. Our small stream runs over its banks fairly often. This happens predictably during the spring melt but also countless other times during the summer when we are visited by heavy rains. Accompanied by our dogs, we walk the length of the stream every afternoon or evening. Even modest flooding alters the landscape of the property. We enjoy seeing the changes that come after a heavy rain and often find ourselves playing explorer both up- and downstream.

Of course, not all of the adventures on the farm are created by nature. Some of what we value also comes from human work and cultivation. We've cleared three of the five acres of burdock and other weeds and have discovered the perfect hammock tree where we can nap or read by the stream. The farm once again has a large garden and rhubarb, raspberry, blackberry, and strawberry patches.

Moving to the farm has changed me. I am more patient and relaxed. While I still work well more than forty hours a week, I have found more time for leisure and appreciation of the natural world. On weekends you would be more likely to find either of us working in the yard or reading by the stream than in the office. Since I produce more of my own food, I also cook more often and more healthfully. I find myself planning meals around what's ripe in the garden or what I have preserved for the winter months. Weekend evenings consist of friends coming over to share in our harvest and sit around the campfire.

Skunk Hallow is not a wilderness area. Wisconsin's wilderness areas are in the far northern parts of the state. An inquisitive and logical reader might wonder what, if anything, a story about moving to an old tobacco farm in Coon Valley, Wisconsin, has to do with wilderness. As an educator I would applaud such good critical thinking skills. I would reply, however, that connection to place, stewardship of land, and valuing one's natural surroundings has everything to do with protecting wilderness areas. A person cannot value wilderness that is "untrammeled" by humans unless she first values what is directly around her. My deep connection to my natural surroundings motivates me to actively seek protection of wild areas far away from me. It is through understanding the workings of the ecosystem where I live that I can respect the need for ecosystems without a human presence. Being a steward of place deepens my connection to land, wildlife, and wildness not only here, but everywhere.

# Chapter Two
# Values and Common Ground

*"He who owns a veteran bur oak owns more than a tree. He owns a historical library, and a reserved seat in the theater of evolution."*

– Aldo Leopold
*A Sand County Almanac*

Wilderness advocates of today have laid out a call to action and argued for citizen activism in appealing for wilderness designations in the lands surrounding them. They are correct to point out that the need for grassroots activism is both imminent and critical. But in order for a person to be motivated to take action, she or he must value wilderness or, at the very least, the natural world.

The word *value* has a multitude of meanings. To some, wilderness is instrumentally valuable in the sense that there is money to be made from harvesting, extracting, or overcoming it. For others the value is intrinsic: wilderness is valuable in and of itself, not tied to a human's ingenuity, pleasure, or profit. Most policy-making involves finding common ground among diverse sets of values.

## Facts and Values

Young adults often challenge my suggestion that policy-making has more to do with values than with facts.

As constituents we want to believe that our elected officials are making decisions based on solid evidence and sound facts. The truth is that facts are neutral, whereas decisions about the course of human actions are not. Imagine a scenario in which there is debate about whether a state should allow mining in a specified area. We might expect that proponents of the mining operation will stress the importance of jobs and the potential growth of the economy, while opponents will focus on the potential environmental costs of mining. Both sides will be armed with facts. For example, studies might confirm that the area contains metallic sulfide deposits, that the mining operation will create one hundred jobs, and that these jobs will give a boost to the local economy. We might also have facts about the heavy metals and other dangerous waste the mine is likely to produce. Hard data might show what this waste can do to both human health and the natural environment. But facts themselves will not tell lawmakers what to decide. Rather, lawmakers must assess what citizens value more: economic growth or a pristine environment.

**Value Definition**

The word *value*, when used in this way, describes a desirable situation. When defining it for others I like to argue that the word *value* is synonymous with the word *goal*. If I value something, then it is my goal to bring it about, protect it, or enforce it. For example, if I say I value honesty, then it is my goal to be honest, and my peers can expect me to be honest. If I claim to value honesty but get caught telling all sorts of lies, those who catch me will suggest that I don't value honesty at all. If I say I value wilderness, people can expect that I will work to protect wilderness.

A twofold problem emerges when assessing values about wilderness, however. First, the wilderness designation by Congress limits human activity in that area. And second, humans have multiple and competing values at the same time.

## Natural Values

People hold a wide range of values about the natural environment. There are those who deem the natural world to be valuable for the raw materials it can provide. These folks are often labeled as cornucopians by environmental philosophers. Cornucopians generally believe that human ingenuity and technology can solve any environmental problems that arise due to resource extraction. While cornucopians are generally not thought of as environmentalists, dividing cornucopians and environmentalists into separate camps is an overly simplistic dichotomy. Environmentalists, too, value many things. Some environmentalists value the environment for what it can do for humans. They may enjoy camping, backpacking, or fly-fishing and value the wild environment as a playground. These individuals are labeled anthropocentric (human-centered) environmentalists. Other environmentalists value plant and animal life and are labeled biocentric (life-centered) environmentalists. Still others value the system (ecocentric) or the biodiversity of life on the planet as a whole as compared to just a particular ecosystem. If all of these groups have slightly different values, then it follows that they also have slightly different goals for wilderness. In deciding whether or not to designate a given wild area as wilderness, Congress must balance all of these goals—no small challenge.

A handful of wilderness advocates argue for the wilderness designation at the expense of any—or even every—other value. After all, the anthropocentric, biocentric, and ecocentric environmentalists would all benefit from a new wilderness designation. However, not all of these individuals would have their values maximized at the same time. Human enjoyment of a wild place can certainly have an impact on other forms of life or on the ecosystem itself. In addition, once an area is designated as wilderness, the kinds and amount of human activity allowed in the area are very limited. These differing values often lead to emotional battles between cornucopians and environmentalists. Too often those in the environmental camps demonize those who would prefer to see wild areas used to collect raw materials for human use. Undoubtedly many of the individuals who argue in favor of logging or mining interests wish to make money. It is unlikely that they intrinsically value the money itself, but they may value other things, such as family, home, or a business for which they need the income. There is nothing inherently evil about making a profit to support oneself or one's family.

## The Tragedy of the Commons

When speaking about balancing values, I ask young adults to engage in a thought experiment roughly based on Garret Hardin's tragedy of the commons. In his essay on this economic theory, Hardin explains how decision-making based on rational self-interest can lead to the destruction of common resources. For our thought experiment, I tell my students that I have decided we should start a commune near Hayward, Wisconsin, a beautiful area near the Chequamegon National Forest. I explain that we are going to do all of the things I imagine people in a commune would do, including growing our own food, making our own clothes, and raising animals. I tell them that I have just enough money to buy each of them five sheep. Twice a year we will shear our sheep and take the wool into Hayward to sell, and each of them will earn twenty dollars for the wool of their five sheep. But what to do with the twenty dollars, as we all have everything we need at the commune? Well, it just so happens that in my thought experiment, additional sheep cost ten dollars. Wouldn't it make sense, I ask each student, for you to buy one or two more sheep? The next time you sell your wool, you can make between twenty-four and twenty-eight dollars. If you buy two more sheep in six months, you could earn up to thirty-six dollars. At that point you could buy three more sheep.

Over time each individual could purchase more and more sheep and in turn increase his or her earning power. Citizens acting in their own rational self-interest should seek to increase their individual wealth. How else will these young adults move away from the commune after they tire of me? The problem, of course, is that if every person behaved in this manner, our commune would soon be overrun with sheep and we would exceed the carrying capacity of the land. This is the message of the tragedy of the commons. In order to avoid it, we must choose to forgo some of our rational self-interest and instead focus on the good of the whole. During this exercise I make the important point that there is nothing wrong with trying to increase your individual wealth. No one is characteristically evil for wanting to do so. The same is true for those who hold more cornucopian or anthropocentric values. There's no reason to demonize those who want to provide jobs or to grow the economy. Most environmentalists also want a strong and healthy economy.

## Value Hierarchies

It is part of the human experience to hold many values at the same time. Some of these values are competing, and as a result humans tend to rank their values in a kind of hierarchy. For example, I value the wildlife, including the rabbits, around our farm. But I value my vegetable garden more than I do the rabbits, so I put up a fence to prevent the rabbits from eating my vegetables. In other words, I give more consideration to my vegetables and the benefits they provide my family than I do to hungry rabbits. Most citizens would say they value wild areas, just as most would say they value the economic well-being of their communities. Nearly everyone would admit to valuing their family and friends. We all probably also value clean air and water, education, national security, health, and independence. The truth is, we share more values than we realize. But our value hierarchies are rarely identical. A person arguing on behalf of a wilderness designation is valuing wilderness over economic growth in that particular case. Likewise, a person arguing on behalf of logging interests is saying that in this case he or she values jobs more than wilderness.

## Common Ground and Shared Values

Even if all citizens agreed that wilderness is valuable and, by extension, deserves protection, we would still need to reach consensus about what constitutes wilderness. Since policy-makers must take values into account, they must consider all of the ways wilderness can be constructed. In a chapter in *The Great New Wilderness Debate* dedicated to cataloging different arguments for wilderness preservation, Michael P. Nelson comes up with thirty of them: natural resources, hunting, pharmaceuticals, the protection of air, wetlands, and species, life-support, physical therapy, recreation, mental health, aesthetics, inspiration, spiritual encounters, scientific experiments, understanding land health, biodiversity, education, understanding evolution, cultural diversity, our natural heritage, self-realization, disease sequestration, sanctuaries from societal structures, the creation of myth, the raw material for civilization, democracy, social bonding, animal welfare, future generations, yet unknown benefits, Gaia, and for itself (intrinsic value). There are likely many more.

Likewise, in *A Sand County Almanac* Leopold discusses the value of wilderness for recreation, science, and wildlife. There are other individuals for whom the definition of wilderness is not about preservation at all, but about management. Those more in tune with a cornucopian point of view may value wilderness for its economic worth. What each individual values about wilderness translates into his or her goals for wilderness use or preservation.

The political definition of wilderness does not account for many of the values listed above. In order to garner broad public support, policy-makers often use a more inclusive conception. This is not to suggest that the political definition should be changed, but rather is an acknowledgment that wilderness advocates employ more inclusive language when working with (and against) other constituencies. These broad coalitions with diverse interests can be much more effective in reaching compromise. One tactic often used to help find common ground is to uncover shared values. Even if individuals have different things at the top of their value hierarchies, this doesn't mean that they don't share values. The only way to know what an individual values is to ask him or her. Finding common values requires conversation and the willingness to compromise on deciding the most important goal or outcome.

## Barriers to Collective Action

Without agreement, individual citizens would work toward unrelated, even contradictory, goals. Even when the common good can be agreed upon, however, many challenges arise as citizens try to work together, ranging from fairly straightforward problems of coordination to much larger setbacks in which social trust or social capital has broken down. There will always be members of society who favor their own interests over the interests of the group or the whole. When too many individuals do this, we again face Garret Hardin's tragedy of the commons. There will also always be individuals who stand to gain from the collective actions of others even though they do not participate. Those who do not work to save wilderness today will still benefit from the existence of wilderness in the future. We generally refer to those who consciously refuse to participate as free riders. These challenges to collective action can often be overcome with appeals to collective rationality and discussion about what it means to be a member of a community.

One collective action problem, however, is much harder to overcome. The challenge, often referred to as the prisoner's dilemma, occurs when one group of citizens either cannot predict what other citizens will do or does not trust the others to keep their word. In game theory the classic prisoner's dilemma describes two individuals (A and B) who have been put in jail for allegedly committing a crime. They are kept in separate cells and do not know what the other will do or if they can trust the other. Each is offered a deal by the police. Further, the police offer clemency if a confession includes turning in the other suspect. The prisoners can confess, or they can stay silent. There are four possible outcomes. If A confesses and B does not, then A will be set free and B will serve three years. If B confesses and A does not, then B will be free and A will serve three years. If both confess, then both A and B will serve two years. If neither confesses, both A and B will spend one year in jail.

Not knowing what the other will do, a logical person should choose to confess even if doing so means betraying the other. While both prisoners staying silent would produce the most optimal outcome for both, making this choice requires knowing what the other person will do and trusting him or her to do it. Lacking that knowledge or trust, the rational thing to do is to act in accordance with your own self-interest—in this case, to confess. Unfortunately, when it comes to environmental dilemmas, and especially to the protection of wild areas, citizens often are not confident about the motives of others or flat out don't trust others. For example, there is a great deal of distrust between those who advocate for drilling in or near a wild area and those who want to see the area protected from human interference.

The only way to surmount these collective action problems is through communication and relationship-building. This requires open channels of communication so that decision processes are transparent and competing groups can come to trust one another.

## Compromise

*Compromise* should not be considered a dirty word. Humans have to compromise on a daily basis. For example, my husband's top choice for a movie is an action film, while mine is a romantic comedy. There are other movie genres we agree on, however; we both like comedies, political dramas, and thrillers. When we're going out to see a movie, I write down my top choices in order, and he writes down his. Then we select the movie that ranked highest on both of our lists. In doing this I forgo my top choice as often as he must forgo his. This certainly doesn't mean we always have to give up seeing the film at the top of the list. We might go see it with a group of friends another day or wait until it comes to Netflix. It may seem overly simplistic, but this method allows for something important: sometimes one of us will list as our top choice a movie that we know the other one wants to see. This usually happens when one of us doesn't feel strongly about any of the films playing, but we know that the other person does feel strongly about one.

What an individual gives up when he or she compromises is referred to by economists as an opportunity cost. This is the cost of any alternative that must be sacrificed in order to engage in a common objective. The word *cost* itself refers to any reward or benefit a person could have received by choosing his or her preferred option. By compromising on my movie choice, I may be giving up the emotions or excitement I was hoping to experience, or I might miss out on discussing the movie with friends who have seen it.

People confront opportunity costs every time they make a decision, from one as simple as what to eat for dinner to something as complex as choosing where to live. Opportunity costs are part of citizens' everyday life. But when it comes to policy-making, these costs are often cast in a negative light. When one group of citizens has to forgo something because another group has a different set of desires, the opportunity costs are often tallied and become a source of conflict. Too often compromise is seen as selling out. But progress can be made when citizens are willing to acknowledge the opportunity costs paid by others. Acknowledging and respecting those costs is a necessity in the process of finding common ground.

When seeking common ground, it is always easiest to start with the values we share. Once the parties involved form a connection, it becomes easier to compromise. Even a process as simple as the one my husband and I use to choose a movie is helpful in the effort to find common ground. For example, two individuals might agree on the need for wilderness to provide biodiversity, even though biodiversity is not on the top of either of their lists of reasons for saving wilderness. Let's say one of these individuals places hunting at the top of the values list; the other ranks backpacking highest. But they acknowledge that they both included biodiversity on their values list. With that shared value in mind, they can work for consensus on times and locations for hunting and times and locations for backpacking. If the hunter identifies a set of weeks during which hunting is best, perhaps the backpacker will recognize that backpacking during those weeks is just not that important to him, as long as it is allowed in the area during most of the year. Likewise, the person who values wilderness for aesthetics and the person who values it for future generations may not be all that far apart in their policy-making desires. Leopold made a case for cooperation between interests in his essay "Coon Valley: An Adventure in Cooperative Conservation," writing, "The stage, in short, is all set for somebody to show that each of the various public interests in land is better off when all cooperate than when all compete with each other."

Finding shared values may not always be possible, however, and it is in these situations that the most compromise is required. In today's political climate compromise is often viewed as synonymous with defeat. But the American political system is designed to enable compromise. No single group will get everything it wants all of the time. If an area is not designated wilderness but still acts as a wildlife corridor, then surrounding wilderness areas will prosper because of it. Those who argue against wilderness designations can be, and in rural areas often are, stewards of the land. Rural ranchers are often suspicious of federal government involvement in land use decisions, especially in the West. Yet their ranches are beneficial to surrounding wilderness areas.

A good example is the area just outside the northern boundary of Yellowstone National Park near Gardiner, Montana. Paradise Valley is home to many cattle ranches that have been operating for generations. Having done policy research in this area, I can attest to the ranchers' strong skepticism toward federal regulations and bureaucrats. Yet the wildlife traveling through the Greater Yellowstone Ecosystem is dependent on these agricultural lands as natural wildlife corridors. Yellowstone is, after all, partially made up of and surrounded by areas with the wilderness designation. There may be some avid environmentalists who would like to see all of Paradise Valley return to wilderness, but most citizens recognize that the ranchers have been raising cattle here for generations and believe they have a right to do so. The truth is, without the ranchers the entire area would likely be a large subdivision or a tourist destination complete with a waterpark. By maintaining their ranches, these residents allow wolves, elk, bison, and other animals to move between wild areas.

Aldo Leopold warned against what he called splitting, advocating for large, contiguous wilderness areas as opposed to smaller, segregated plots of land. Large areas are better at preserving ecosystems and habitat for large prey animals. But in the case of Yellowstone, the human presence has been there for generations and serves a very important purpose. Moreover, the ranchers are stewards of the land. By virtue of having ranched there for generations, they have demonstrated that they can both work the land and take care of it.

One common criticism of cattle operations in the West is related to grazing. While detailing the entire grazing controversy is not within the scope of this book, its role regarding wilderness protection does deserve consideration. As the West was being settled under the Homestead Act, those willing to work the land and, according to the philosophy of the day, increase its value for a certain number of years were able to gain ownership of the land for little to no cost. Since the western United States is water deprived, many of the settlers homesteaded plots of land near creeks, rivers, and other water rich areas, leaving large plots of undeveloped land. As time passed, the federal and state governments assumed ownership of the undeveloped land. This created a checkerboard pattern of private and government-owned land in many states in the West.

In order to make a living in the arid West, many settlers took to cattle ranching. Vast acreage is needed for grazing in the West because the dry conditions limit plant growth, and thus ranchers were allowed—and in many cases encouraged—to graze their cattle on public lands. It was not until the Taylor Grazing Act of 1934 that ranchers were asked to pay for their use of public lands, and at that time the price reflected only the cost of maintaining the land. The Taylor Grazing Act and subsequent legislation also offered ranchers some planning security for their ranching operations via the purchase of grazing permits. These permits operate much like leases, guaranteeing that ranchers can graze a certain number of cattle for a certain number of years on public lands.

What makes grazing controversial today is complex. While some environmentalists worry about overgrazing of the grasslands, others worry about the damage cattle do to riparian zones. The costs of grazing permits for public lands is currently much lower than the market value of permits for grazing on privately held lands: $1.35 Animal Unit Month (the cost of grazing one cow or five goats or sheep for one month) on public lands compared to $10 to $15 AUM on private lands in many western states. Of course, ranchers require grazing for only part of the year, but the costs add up. For example, grazing a herd of three hundred cattle on public lands for the three summer months costs approximately $1,215, compared to $9,000 to $13,500 on private lands.

Ranchers currently can secure grazing permits in national forests, which often include and surround wild areas. Wilderness advocates have targeted the reallocation of grazing permits near both wild and wilderness areas, charging that cattle cause environmental degradation. Environmental advocacy groups have petitioned the US Forest Service to allow competition for the grazing permits, which would allow environmental groups to purchase the permits in place of ranchers.

Grazing cattle on public lands has been a necessity for ranchers in the West for over a century. Many of them could not operate at capacity without the use of public lands. The ranchers I have met just outside of Yellowstone near Gallatin National Forest have been operating in the area for generations. For them, ranching is more than a job or even a way of life, it is an identity. Despite the fact that they live in one of the most beautiful areas in the United States, they struggle to keep up with the costs of ranching. Many of these ranchers' way of life would be threatened without the current system of grazing permits. It is not practical to expect ranchers who are dependent on public lands to withstand an increase in grazing costs by 900 percent. These ranches also preserve open space and wildlife corridors. Uncovering the common good will require finding common ground in a very contentious debate. But compromise is crucial. If the public prefers to see open land as opposed to Starbucks in Paradise Valley, then perhaps grazing permits in the national forest there are a necessity. In our endeavor to save wild areas, we should listen to Voltaire's advice and not let the perfect be the enemy of the good.

When finding common ground is not possible, elected officials can ask questions to help them determine whether or not a given area should be designated as wilderness. The first is about the uniqueness of the area. If an area is especially distinctive—offering landscapes found nowhere else, or providing habitat for a rare or threatened species, for example—then it might be the uniqueness itself that is of value. Another question is whether leaving the area open for human activity is of vital importance for community survival. If so, then perhaps the needs of the community are more valuable than the wilderness. However, elected officials must guard against the inclination to think of wildlands solely in terms of profit, the economy, or jobs. Rather, they should seek to balance as many values as possible. When confronting a controversial decision about whether or not a given area should be designated as wilderness, elected officials must question whether leaving the area open for human activity will solve a problem for the humans in the community.

In a chapter called "Feeding People versus Saving Nature?" in *World Hunger and Morality*, Holmes Rolston argues that both short- and long-term effects should be taken into account when choosing between humans and nature. It sounds shortsighted to argue that there are times when wild things should be valued more than human beings. However, sometimes this is the case. Using Rolston's example, if opening a nature preserve to a growing population will provide only a temporary fix to the problem of hunger, then opening the nature preserve should be considered very carefully. If in ten years there will be no nature preserve and there will still be hunger, then the preserve should not be opened, even if people are starving today. When balancing the needs and values of a community, elected officials should ask similar questions, considering, for example, whether a wilderness area rich in minerals could provide long-term stability to the surrounding communities or only a short-term fix to economic woes. In Leopold's words:

Sometimes I think that ideas, like men, can become dictators. We Americans have so far escaped regimentation by our rulers, but have we escaped regimentation by our own ideas? I doubt if there exists today a more complete regimentation of the human mind than that accomplished by our self-imposed doctrine of ruthless utilitarianism. The saving grace of democracy is that we fastened this yolk on our own necks, and we can cast it off when we want to, without severing the neck. Conservation is perhaps one of the many squirmings which foreshadow this act of self-liberation.

## Vignette
# Lost in the Wilderness

One of the first wilderness areas I remember visiting is the Boundary Waters Canoe Area Wilderness (BWCAW) of northern Minnesota and Ontario, Canada. I was on a youth outdoor adventure trip with an itinerary that included canoeing into the Canadian waters to set up camp. Unfortunately, we lost a person on this trip. To be more precise, he got lost, and we had to get some help to find him. For the purposes of this account I will call him Pastor G.

The BWCAW is a wilderness area made up of connected waterways, lakes, and bogs. Our group intended to canoe several miles into the wilderness before setting up camp for the week. I remember vividly the day we picked up our canoes and supplies from an outfitter in Ely, Minnesota. We spent the night in cabins in Ely before embarking on our adventure the next morning. At morning light we diligently canoed through the last area in which motorboats are allowed and headed into the wilderness. I no longer remember how many miles we canoed or even the name of the lake that was to be our destination. I do remember, however, that during one portage we saw a note that read "Bear seen here," dated the previous day. This made me anxious in a way I had never experienced before. But as it turned out, bears would not be the problem on this trip.

Since our group was large, we split into two groups and set up camps about a half mile apart along the shore. We hung our food to keep it out of the reach of bears and then set about doing all of the things a group of young campers might be expected to do, including swimming, starting a fire, and playing cards.

A few days into the trip, a friend and I decided to hike over to the other camp. Logic suggested that as long as we stayed in sight of the shoreline, we could not get lost. Although we reached our destination safely, when we arrived we told the other group how our logic had been flawed: the steep and rocky shoreline had made it impossible for us to see the lake the entire length of our hike. Luckily my friend was pretty good with a compass, and we were able to go around, as opposed to over, some of the rocky structures. If asked to explain how we accomplished this hike, though, I would have cited dumb luck. One of the members of the other camping party gave us a canoe ride back to our own camp.

The day after our little adventure, Pastor G decided he would replicate our journey (in reverse) and hike over to see us. But while he was attempting to navigate the rocky shoreline, he got lost. He was breaking one of our group's rules by traveling alone; our camp didn't know he was coming, and therefore we were not alarmed when he hadn't shown up four hours after he headed out. It wasn't until later that day, when one of the members of his camping party came to collect him via canoe, that we knew he was missing. Both camping parties then set out to find him. We looked until dark with no luck. At daybreak two of the most able canoers set out to get help. Our adult leaders helped the rest of us pack up and head out. As I look back at this story now, as an adult myself, I can't help but wonder why the rest of us didn't stay put. If Pastor G had been able to find his way to either camp we would all have been gone.

As soon as we crossed into the United States we were given a status update. Authorities had already used a search plane to comb the area in which we had camped, with no sign of Pastor G. A second plane was about to begin a search. We canoed as far as motorboats were allowed and were picked up and taken back to Ely. As we rode into town to get a warm meal, I felt sick to my stomach, thinking myself partially responsible for what was happening. I reasoned that if I had not attempted this hike with my friend, then Pastor G would not have attempted it either. I pictured Pastor G lost in the wilderness, his glasses broken, pursued by hungry bears.

When we got back from town, Pastor G was standing next to the shower building. He looked a little roughed up, but he was alive. In fact, he was just fine. To his credit, Pastor G did all of the right things once he determined he was lost. He had ended up at a different lake and decided to stay put and set up camp. Luckily he had some matches and was able to build a little lean-to and start a fire. The next day, when he heard the planes, he went to work to assemble a flag out of his undergarments. Although authorities spotted him from the second plane, they were unsure about the possibilities for landing on the lake, and they went back to headquarters to create a plan. They were eventually able to land a third plane on the lake and rescue Pastor G.

A few years later I returned to that spot with members of both my and Pastor G's families. Our adventure party canoed into the original campsite, and this time, armed with both maps and compasses, the group successfully hiked in to where he was found. I'm not sure what made him want to return there, but I know why I needed to go back. I wanted to trade out my fear of this wilderness area with an appreciation for it. By that time I was planning to go to graduate school to study environmental philosophy, and I was starting to ask deep questions about what comprises wilderness and about the relationship between humans and wilderness. I needed to think about what it means for a human to be lost in the wilderness.

Perhaps one of the most important lessons I've drawn from this experience is that we can be more successful as a group when we work together and carry our maps and compasses. To be sure, finding common ground in environmental disputes can feel like being lost in the wilderness. We hold tightly to our own values and often fear that compromise means letting go of our ideals. Without ideals we don't always know what we are looking for. Sometimes protecting wilderness means working with those outside of our own political camp. By working together, we can both find a solution and keep an eye out for bears.

# Chapter Three
# Human Presence, Conservation, and Wise Use

*"Wilderness is a resource, not only in the physical sense of the raw materials it contains, but also in the sense of a distinctive environment which may, if rightly used, yield certain social values."*

– Aldo Leopold
"Wilderness as a Form of Land Use"

Chapter 2 discussed values, but it largely ignored the larger philosophical debate about wilderness. It is worth a brief detour into the world of philosophical inquiry, as our personal philosophies toward the natural world underlie both the values we assign to nature and the political process of determining what is "wilderness." The central question is whether wilderness should be considered "part of" or "apart from" the human enterprise. Our answer to that question directly affects the types of policies we enact for wilderness protection.

**Western and Non-western Conceptions**

The political definition of wilderness includes the assumption that wilderness is "apart from" humans. According to the Wilderness Act, wilderness areas are to be "untrammeled" by humans. While humans may visit, they must not interfere. This position has been rightly criticized as overly ethnocentric and western.

To be sure, the American spirit is one of conquering the wild and untamed lands of the frontier. Westward expansion sought to settle—even "tame"—the West. In the process, prairie lands were plowed under, forests were cut down for lumber, bison were slaughtered, and other species, such as the gray wolf, were eradicated. Nature was seen as something to be conquered, and conquering the land was what made it valuable. This approach was fueled in part by a Judeo-Christian conception of a world created for human use and in which natural objects held no special meaning or life force. Moreover, according to this Judeo-Christian worldview, the natural world was specifically created for human use.

In the United States, wilderness is woven in to our national identity. The founding fathers admired the wildlands of the young republic and placed great value on its wealth of natural resources. At the same time, the immense wilderness was something to be feared and, in turn, conquered. When President Thomas Jefferson commissioned the Lewis and Clark expedition, the explorers' primary focus was to map the territory and create a trade network with Native American tribes. Yet they were also tasked with studying and documenting the plants, wildlife, and geography they encountered. While property is often listed alongside such other core American values as "life, liberty, and the pursuit of happiness," it is clear that natural places and their scientific value were also important to the founders. In many ways the story of wilderness in America is the story of citizens fulfilling some essential human need. It was through overcoming, taming, and working wildlands that citizens proved their character. The emerging American identity included self-reliance, hard work, and deep love of the land.

Eventually there came an acknowledgment that humans had perhaps gone too far in their exploitation of the land and that wild areas needed to be preserved for the future. The Wilderness Act sought to remedy these oversteps by setting aside and protecting wild areas from human interference. In doing so, however, it maintained the human/wilderness dichotomy—the sense of humans being apart from the natural world.

Yet there were non-western views of wilderness prevalent on this land long before the European settlers headed west. Although there is no single shared view of the world from a Native perspective, generally speaking Native groups believe that humans are "part of" the natural world. Wilderness does not exist to be conquered or overcome. Rather, humans are meant to live in harmony with their natural surroundings. This represents a union of the physical and spiritual worlds and of the natural and the supernatural. In the truest sense, this worldview recognizes the interconnection between all things past, present, and future. In an essay written in 1933, Oglala Sioux chief Luther Standing Bear reminds readers about the native view of the physical world:

> Knowledge was inherent in all things. The world was a library and its books were stones, leaves, grass, brooks, and the birds and animals that

> shared, alike with us, the storms and blessings of the earth. We learned to do what only the student of nature ever learns, and that was to feel beauty.

Rather than attempting to overcome wilderness, indigenous peoples around the globe have embraced, respected, and revered it. Of course, like all humans, Native groups were dependent on the world around them. Since as humans they were "part of" nature, however, when they took from the natural world they gave thanks and homage back to it. These philosophical differences shaped how Europeans both viewed and treated Native Americans. While there are stories of explorers utilizing Native knowledge of a particular landscape, the historical narrative more often recorded European impressions of "wild savages" who did not have a desire to learn western cultural ideals. Sadly, this perception led to forced assimilation and the sequestration of Native culture to reservations.

## Conservation versus Sustainable Development

There is a similar divide today between those who want to preserve nature as it is "apart from" human influence and those who believe humans can live and work in wilderness as long as they use sustainable practices. The first group might be called conservationists, while the second adopts the principles of wise use or sustainable development. These philosophical camps envision very different outcomes for designated wilderness.

The conservationist philosophy of today echoes the call of the preservationists of times past. Wilderness should be set aside and preserved for the future. There should not be any human interference because once wild areas are exposed to human exploits, by definition they cease to be wilderness. The roots of this philosophy can be traced back to early environmental authors including Ralph Waldo Emerson, Henry David Thoreau, and John Muir. As Muir wrote,

> Wilderness is not only a haven for native plants and animals but it is also a refuge from society. It's a place to go to hear the wind and little else, see the stars and the galaxies, smell the pine trees, feel the cold water, touch the sky and the ground at the same time, listen to coyotes, eat the fresh snow, walk across the desert sands, and realize why it's good to go outside of the city and the suburbs.

The wise use philosophy, on the other hand, accepts human presence as a natural part of wilderness management. The movement toward sustainable development is embraced not only by cornucopians but also by a subset of environmentalists. Gifford Pinchot, who in 1905 became the first chief of the US Forest Service, believed that conservation and development were synonymous. For example, Pinchot believed that the only way to prevent or reverse uncontrolled forest fires was to manage the land via logging. Likewise, some environmental philosophers today believe that sustainable human activity is an inherent part of human presence in wilderness areas. While some agree there is a time and place for biological preserves, others argue that the concept of pristine wilderness is detrimental to the movement to save wilderness. This ideal, they argue, is flawed in that it rarely exists; holding the ideal causes citizens to devalue land that is used and managed sustainably but still has all of the characteristics of wilderness. Moreover, they argue, the concept of wilderness belongs to developed nations. Developing countries, dependent on their natural resources for economic growth, do not have the luxury of setting aside wilderness for nonhuman interference. A more inclusive definition, this camp argues, would be one that is more global in nature. Fostering sustainable development in wilderness areas abroad requires fostering sustainable development in wilderness areas at home.

Both sides of the conservation versus sustainable development debate have appealed to the writings of Aldo Leopold. And indeed, both philosophies can be found in his writings. The concept of separation between humans and nature is evident in many of his works. *A Sand County Almanac* includes: "Wilderness is the raw material out of which man has hammered the artifact called civilization." And in an essay in *The River of the Mother of God and Other Essays*, he writes, "By 'wilderness' I mean a continuous stretch of country preserved in its natural state, open to lawful hunting and fishing, big enough to absorb two weeks' pack trip, and kept devoid of roads, artificial trails, cottages, or other works of man." These entries certainly seem to support the preservation of wilderness "apart from" human interference.

The vast majority of Leopold's essays, however, are about conservation in some shape or form. In "The

Farmer as a Conservationist" (found in *For the Health of the Land*) he penned, "When the land does well for its owner, and the owner does well by his land, when both end up better by reason of their partnership, we have conservation. When one or the other grows poorer, we do not." After all, much of *A Sand County Almanac* is about maintaining the land in the sand counties of Wisconsin, where Leopold taught farmers sustainable agricultural practices. The land ethic, at its core, is a mandate for human management, not a directive for grouse, deer, or black bears.

Can a person hold both of these philosophies at the same time? Those who advocate for wise use tend to try too hard to mesh Native beliefs about nature with the sustainable use movement. Yet Native Americans don't automatically adopt wise use principles. For example, Native American groups in Wisconsin banded together to stop the development of a metallic sulfide mine near the town of Crandon and the Mole Lake (Ojibwe) Reservation; in another example, the Gwich'in people of Alaska have been vocally opposed to opening the Arctic National Wildlife Refuge for drilling because they are dependent on the caribou migration there. To be certain, there also are Native groups that support sustainable development. Some of the same tribes that opposed the Crandon mine have sustainably logged their lands for generations. The land use debate has more to do with applied values than it does with an underlying nature philosophy.

Philosophical ideals underlie the wilderness debate, but they are not guides to action. As I traverse the controversies of environmental policy today, I have a warm place in my heart for environmental philosophy. It has helped me think deeply about humans' relationship with the natural world. Simply understanding that divergent philosophies exist is an important tool in advocating for wilderness protection. An appreciation of these differing worldviews can help us all navigate the sometimes choppy waters of environmental debates and get to common ground.

## Vignette
# Bear Country

When I visit wild areas, I consider myself to be in the home of the wild animals that live there, and I work to respect them. I hate hearing stories about a bear or other wild animal having to be destroyed when humans' careless or inappropriate actions led to the animal harming someone. And while I understand that authorities sometimes have to remove these animals, I feel strongly that if I am ever harmed by a bear, I do not want that animal to be harmed. All this being said, I have a healthy fear of bears, especially grizzly bears.

It might seem odd, then, that just a few days after our engagement Jeremy and I decided to go backpacking in grizzly country. We had traveled to Havre, Montana, for Jeremy's family reunion and were on our way to stay at a family cabin in Hungry Horse, just outside the west entrance to Glacier National Park. We stopped at a rangers' station in Kalispell to ask about a remote location for backpacking and get the necessary permits. The ranger recommended an area in the Flathead National Forest that includes the Bob Marshall, Great Bear, and Mission Mountains Wilderness Areas.

To get to our first (and what was to be our only) campsite, we hiked about eight miles straight up a mountain. One of the benefits of being newly engaged was that Jeremy carried the pack, while I was in charge of the carrying the map. We stopped at our designated campsite near a beautiful lake. The only other people we saw that day were a few travelers headed out on horseback. At the rangers' station we had overheard mention of someone having an encounter with a grizzly bear mother. The tale included bear spray and the eventual retreat of the bear. With bears on my mind, we bear-proofed our site well, hanging our food and toiletries and locating our tent far from both the food and fire pit. Although we did have a campfire that night, we did not cook as we were still tired from the hike.

While we were putting up the tent I asked Jeremy to practice what we would do if a bear came in the night. It was at that point that we opened the package of bear spray we had brought with us. (I can now acknowledge it would have been smarter to open before we ventured into bear country.) The canister was empty. We created a new plan minus the bear spray and settled in for the evening.

I didn't know what time it was when I woke to the sound of Jeremy's heart beating like a bass drum. Then I heard the cause of his alarm: a loud crash in the bushes, then another. Then I heard sniffing—very loud and heavy sniffing. Something was smelling the area where our food and toiletries were hung. The sniffing noise progressed over to the fire pit and eventually made its way up to our tent. As we tried to stay silent, the animal sniffed around our tent for a minute and then sauntered off in the direction of the lake.

We didn't see the animal, but we didn't have to. The morning light revealed paw prints in the moist ground the size of small dinner plates. They had been made by a black bear or grizzly bear. Curiously, there was also a line in the sand from where our items were hanging toward the fire pit and then over to our tent. I could only assume that was where a large nose had been pressed against the ground. While I consider myself an outdoorsperson, the incident shook me. That bear must have known there was food in the hanging bag. It was summer, so the bear probably wasn't desperately hungry, but I wanted to get out of there. We packed up and made it back to the car without further incident, other than an elk bounding across the trail in front of us and nearly giving me a heart attack. It was only after we were tucked safely back in the car that Jeremy confessed to having huckleberry lip balm in his pocket while we were in the tent. What? It says right on the tube not to use it in bear country!

Confronting the political process can be a bit like confronting a grizzly bear. You can know all of the rules and theories about animal (or political) encounters but still not be able to predict exactly what is going to happen. You can do everything you are supposed to do and still be shaken up. You might even see other individuals who don't follow the rules, who don't take the right steps, and who don't respect the process yet walk out unscathed. Much like taking the right precautions in bear country, successfully creating new public policy necessitates both diligence and responding to what is going on around you. It can require quick thinking and a survivalist attitude.

## Chapter Four
# Wilderness and the Political Process

*"An ethic, ecologically, is a limitation on freedom of action in the struggle for existence. An ethic, philosophically, is a differentiation of social from anti-social conduct. These are two definitions of one thing. The thing has its origin in the tendency of interdependent individuals or groups to evolve modes of co-operation. The ecologist calls these symbioses. Politics and economics are advanced symbioses in which the original free-for-all competition has been replaced, in part, by co-operative mechanisms with an ethical content."*

– Aldo Leopold
*A Sand County Almanac*

The most important aspect of the Wilderness Act is that it exists. Without it we would have no designated wilderness areas. Getting the Wilderness Act passed took decades of work by advocates and an eight-year debate in Congress. And the challenge is ongoing, as getting wilderness areas designated by Congress is a tedious process, requiring a variety of federal agencies to submit reports about the condition of wildlands. Although requiring congressional approval for both the creation and removal of the wilderness designation was born out of political compromise, this restriction has turned out to be an advantage for wilderness advocates.

Congress has removed the designation only a handful times, and in only one year since 1965 has the total number of acres designated as wilderness decreased (with 222 acres of Washington State's Olympic Wilderness removed from designation in 2012).

To be sure, wilderness has benefited from the status quo—the current state of affairs. When trying to advance legislation, the closer the bill is to the status quo, the more likely it is to be adopted. Since the Wilderness Act exists, and since Congress consistently designates wilderness and rarely removes it, proponents of wilderness are in a stronger position than those who oppose it simply because their cause is closer to the status quo.

Eight years might seem like a long time for a bill to be debated in Congress. While political pundits like to lament the pace at which Congress works, the truth is that our political system is designed so that most policy change is both slow and incremental. The balance of power ensures that the fleeting passions of a few cannot overtake the rationality of the majority. I tell young people that I would be more skeptical of any act that passed swiftly and without debate than I would be of one that took eight years to pass. There are legitimate debates about the best use of public lands, and if citizens are being properly represented by those they elected, these debates should play out in the halls of Congress.

Decisions about land use can be especially controversial because so often humans identify personally with the lands they live near and work on. This, along with the diverse values attached to the natural world, almost guaranteed that there would be a lively policy debate about the proper way to protect wilderness. The process of bringing an idea to Congress and creating a law often takes far longer than eight years. The public must first be aware of and support the legislation so that they put pressure on lawmakers to consider the issue. An issue that has captured the public's attention is called a salient issue. By 1964, the year the Wilderness Act was made law, wilderness advocates had already been working for decades to make the issue of wilderness protection salient. In fact, Leopold was calling for the government protection of wilderness almost forty years before the act was passed:

> If the wilderness is to be perpetuated at all, it must be in areas exclusively dedicated to that purpose. . . . Like parks and playgrounds and other "useless" things, any system of wilderness areas would have to be owned and held for public use by the Government. The fortunate thing is that the Government already owns enough of them, scattered here and there in the poorer and rougher parts of National Forests and National Parks, to make a very good start. The one thing needful is for the Government to draw a line around each one and say: "This is wilderness, and wilderness it shall remain."

## The Policy-making Process

Many Americans would recognize the lyric "I'm just a bill. Yes, I am only a bill." This *Schoolhouse Rock!* cartoon that first aired in 1974 depicts how a lowly bill becomes a law. But of course this three-minute rendition for children presents an oversimplified version of the process by which a bill makes its way through Congress. And it doesn't even attempt to describe the efforts that lead to that point. Policy-making requires the work of citizens. First, someone must decide that a problem exists. Next, that person must try to convince others. This may require getting others to sign a petition or going to see an elected official to ask him or her to do something about the problem.

Even when a bill finds a sponsor in the House or Senate and is introduced in Congress, it is unlikely to become a law. On average almost six thousand bills are introduced in each session; roughly three hundred of those make it through the process. The rest get lost or are killed in committee. For example, part of the reason grazing fees have stayed low on public lands compared to private lands is that key western congressional members have made sure to position themselves on the oversight committees for grazing.

Even if a bill makes it through one house, it must be approved by the other house, including its committees. Finally, the president must sign the bill.

The problem (real or perceived) that citizens seek to solve also must be tied to a solution—what should government do about the problem. After all, there are only so many things government can do. To solve a given problem, elected officials can choose to regulate, educate, charge a fee, subsidize, and so on. There are likely to be more differing opinions about what the government should do about the problem than about the problem itself. In the case of the Wilderness Act, the shrinking of America's wildlands was defined as the problem. The solution entailed the federal government designating lands on which there is to be no human interference. Considering Americans' love of personal liberty, our dependence on natural resources, and the enormity of the problem, eight years doesn't seem all that long.

## Congressional Designation

The authors of the Wilderness Act believed that wilderness needed congressional protection. It might seem like it would be more efficient to allow a director in the National Park Service or the US Forest Service to designate wilderness areas. In fact, this was already happening before the Wilderness Act was passed. Such a process would circumvent the political skirmishes in Washington, DC. There are a couple of potential problems with agency designation of wilderness, however. The first is the fear of regulatory capture, which occurs when special interests secure the loyalty of a regulatory or administrative entity. A logging company executive, for example, might have close ties to a regional director in the forest service due to the longevity of their work history together. If left in the hands of the agency administrator, this type of influence could result in the redrawing of wilderness boundaries to benefit a special interest. A second problem is that administrative appointees are not directly accountable to the public via elections. If the public was dissatisfied with administrative decisions about wilderness areas, they would not have a mechanism for removing the administrator.

An even more significant problem with agency designation is that the directors in such agencies already work under mandates that have the potential to conflict with wilderness preservation. The Roosevelt Arch at the north entrance to Yellowstone reads "For the Benefit and Enjoyment of the People." While some of Yellowstone can be managed as wilderness, it cannot all be wilderness. More than 3 million people visit each year, requiring the maintenance of roads, lodges, and concessions. Bringing the national park experience to the masses is not exactly congruent with managing the land for minimal human impact; a park superintendent must manage for both needs. Letting Congress decide where wilderness boundaries are located lessens the decision-making burden and potential conflict of interest for park administrators.

A number of agencies are tasked with submitting potential wilderness areas to Congress, including the Bureau of Land Management (BLM), National Park Service (NPS), US Fish and Wildlife Service (USFWS), and US Forest Service (USFS). Each reports to Congress every year on any areas within its jurisdiction that fit the political definition of wilderness. To do so, most of these agencies engage in a planning process that includes citizens, holding public meetings, and soliciting written feedback. While meetings are subject to being highly localized events, the vast majority of public documents and the public comment process can be found online. Citizens can read agencies' planning documents via the federal register and can offer comments from their home computer.

## Partisanship

The airwaves today are filled with supercharged partisan bickering over almost every policy issue. Traditionally, however, the fight for wilderness has enjoyed bipartisan support. The Wilderness Act did an amazing thing by clearing the way for Congress to determine wilderness designations. It defined the problem and tied it to a preferred solution. In other words, Congress does not have to debate whether wilderness is important or whether it should be preserved. Rather, Congress has to decide only whether or not any suggested area fits the definition of wilderness.

While this should be a largely nonpartisan issue, there are challenges from some groups, such as the Tea Party, that argue for less government intervention in the lives of citizens. Partisanship also surfaces in debates about the kinds of activities that should be allowed in and around designated wilderness. As noted in Chapter 1, one of the recurring disputes is the debate over drilling in the Arctic National Wildlife Refuge. When vice presidential candidate Sarah Palin uttered the phrase "Drill, baby, drill" during the 2008 vice presidential debate, she was repeating what had essentially become the Republican campaign slogan. While Democrats have continuously voted to support alternative energy solutions, many Republicans have stressed the need for domestic oil and gas production. But even debates about wilderness in areas that possess these natural resources need not be delineated as partisan.

It is important to note that the ANWR controversy, which appears to pit Republicans against Democrats, is not necessarily representative of a deeper ideological divide when it comes to the environment. Traditionally the Republican Party has supported environmental protection. President Theodore Roosevelt was both a

Republican and an ardent conservationist. His love of nature led the way to the creation of the National Park Service. Republican President Richard Nixon signed legislation that created the Environmental Protection Agency in 1970. (Although to be fair, Nixon also vetoed the Clean Water Act; Congress overrode his veto.) Public opinion polls show that members of both parties care deeply about the environment. The source of disagreement in modern politics seems to be more about the necessary methods for protecting the environment as opposed to caring about the environment itself.

I run an experiment in my environmental policy classes to demonstrate that concern for the environment is not necessarily tied to an individual's political ideological perspective. I ask two students to leave the room. If I know them well enough, I try to choose one who has a more politically conservative ideology and one with a more politically liberal one, so as not to bias the results. Then I hand a notecard to the students remaining in the room and ask them to rank their concern for the environment from 1 to 10, with 1 indicating no concern (i.e., "I drive an SUV around the city to prove I am not concerned about energy consumption or air quality") and 10 showing grave concern (i.e., "I keep myself awake at night thinking about all of the environmental problems our country is facing"). Then I tell them to turn the cards over and rank their own political ideology from 1 to 10, with 1 indicating extremely conservative and 10 indicating extremely liberal. I tell them that when the two students come back in from the hall, I want the participants in the room to reveal only the side of the card that indicates their concern for the environment. Once everyone is back together, I draw an imaginary line down the center of the room to represent the political spectrum: one end represents an extremely conservative political ideology; the other end represents an extremely liberal political ideology. The two people who left the room must agree where to place each of their classmates on that line based only on the number indicating their concern for the environment (remember, the young people who left the room don't know the other side of the card has been marked).

The students typically place those with a low number of concern for the environment (1–3) on the extremely conservative end of the political spectrum and those with a high level of concern (7–10) on the extremely liberal end. Once every participant is placed, I ask everyone to turn their cards over. If partisan stereotypes were accurate, we would see a correlation between the numbers for concern and ideology (those who indicated 1 for "not concerned" clustered with those who chose 1 for "extremely conservative"; those who marked 10 for both "extremely concerned" and "liberal" similarly clustered). However, in all but one semester the results have been a completely mixed bag. The majority of young people are moderate on both scales, but even at the ends of the spectrum the stereotypes don't hold up. I see self-described liberals who are seemingly unconcerned about the environment, and self-described conservatives who are very concerned. In fact, recently I saw the biggest difference in the numbers yet. Two young people whose political ideology was decidedly conservative (one had selected 2 and the other 3 for their political ideology) had both selected 9 for their level of concern for the environment.

Many of the young adults who attend the university where I work are from rural agricultural areas. I have found again and again that the vast majority of farmers and ranchers are tremendous stewards of the land. They also tend to hold a more politically conservative viewpoint than their urban counterparts. The two young men with the surprising numbers were both from rural farming families. Not surprisingly, over the course of the semester I learned that they tended to favor a market approach, as opposed to government regulation, to solve environmental problems. But the larger point here is about shared values. Over the course of the semester these two conservative young men revealed that they valued the natural world no less—and in many cases more—than did their liberal classmates.

When we can start from a place of shared values, political compromise is possible. Political pundits tell us that if we are "Republican" we must value or believe "y," and if we are "Democrat" then we must value or believe "z." But the truth is that we have a lot more shared values than we do disparate ones. Moreover, the environment should not be a partisan ideological issue. People on both ends of the political spectrum value the environment and wilderness. We may have honest disagreements about land use, but these should not be determined simply by our political affiliations.

Our debates about wilderness should contain a much deeper discussion of values. As citizens we must seek to understand land use disagreements from the viewpoint of those affected by the decisions. When all persons in a debate feel like their side has been heard, it is much easier to find common ground. Appeals to partisanship only shut down the dialogue and render compromise as defeat. Remembering the history of the Wilderness Act and its bipartisan support should inspire wilderness advocates of today to listen to those on the other side of the aisle. Since creating new wilderness areas as well as maintaining the current ones requires the cooperation of members of Congress, citizens should work together to make suggestions that members of Congress can support as bipartisan.

## Vignette

# Bison, Brucellosis, and Values

Growing up in Wisconsin, I certainly had heard of the American bison, but I can't say I learned much about them. In school I learned about the mass killing of the bison across the American West and the recovery of the species in Yellowstone National Park, and of course I recognized bison as an iconic American symbol, but I wasn't all that interested in their plight. That changed in the summer of 2009 when I enrolled in a faculty class called the Stewardship of Public Lands offered through the American Association of State Colleges and Universities (AASCU) and taught by the Yellowstone Association. The purpose of the course was to bring together a group of college faculty, diverse in both discipline and location, to study public land controversies in and surrounding Yellowstone National Park. While there we discussed these controversies, the costs for various stakeholders, and the best opportunities for finding common ground.

The seminar was a life-changing experience. It changed the trajectory of my work, but more importantly it taught me an important lesson that I thought I already had learned.

During the course I learned about brucellosis, a bacterial disease that causes bison and other ungulates to abort their first calf. Most bison infected with brucellosis go on to live normal, healthy bison lives, and the National Park Service has traditionally not managed the disease, its philosophy being to let nature take its course. This would not be a problem if it weren't for the fact that bison migrate out of the park during snowy winters. The ranchers who surround the northern and western entrances of the park have good reason to worry about their cattle contracting brucellosis from infected bison. Until 2011 ranchers detecting the disease via the blood testing of their herds were required to kill the whole herd if even one instance was found, and states that found more than two instances lost their brucellosis-free status, meaning they could no longer transport cattle across state lines. Even after regulations were loosened in 2011, ranchers surrounding the park had to battle the perception that their cattle might be diseased, causing their cattle's market value to drop. A settlement among four federal agencies and two state agencies resulted in a management plan. Bison that left the park were to be hazed back. Those that did not return would be tested for brucellosis, and those testing positive could be sent to slaughter. When the population of bison in the park was 3,000, however, all bison that left could be killed. During the especially harsh winter of 2007–2008, the population of bison in Yellowstone dropped from 4,500 to only 2,500, with about 1,800 of them being killed for leaving the park.

As one might guess, this problem involved a lot of stakeholders: park visitors, park administrators, park scientists, other federal offices, state offices, animal rights organizations, community residents, ranchers, and advocacy organizations to name just a few. During the seminar we spoke to these stakeholder groups. All were impassioned, and all were affected in very different ways.

I fill the political science and public administration classes I teach with reminders of the role values play in policy-making. I talk about varying stakeholders and work hard to be as neutral as I can be when it comes to environmental political controversies. I am well aware of the benefits of experiential learning and sending young people into the real world. Yet as much as I intellectually understand these things, something changed for me during these conversations with Yellowstone stakeholders. As I listened to their varied stories, I found that I agreed with all of them, even those that were exactly contrary to one another. I empathized with all of them and cared about all of them. I suddenly understood citizenship and political disagreement in a way I had not before.

It was during that week that I truly learned in practice about values and policy-making. I had nothing at stake in the brucellosis controversy, so it was easy for me respect differences of opinion, to adopt values on all sides of the debate, and to understand that all those values are important to the policy-making processes. The brucellosis stakeholders shared certain values, but those values were not necessarily placed in the same hierarchy. I learned that both one of the ranching families and members of a local environmental group, often seen as opposing forces, had written letters to the state calling for an end of the depopulation mandate for whole herds. Of course, the rancher valued both the cattle and her livelihood, and the environmentalists valued free-roaming bison, but they wanted the same policy outcome. During that week I came to understand how important it is not just to tell young people about varying opinions but to expose them to the ways these ideas play out in the real world.

It is easy to become so entrenched in our own value systems that we fail to see the whole picture. As hard as I work to be neutral in the classroom, I also get caught in my own set of values. Before my Yellowstone experience I gave myself credit for understanding the rural value system. After all, I grew up in a rural community and am still tied to its agricultural heritage. My father's parents were farmers. I have seen firsthand how those in the farming and ranching communities are stewards of the land. During my college years I became an environmentalist, and my value priorities shifted toward protection of the environment at all costs. At Yellowstone I realized that talking directly to the stakeholders and understanding their value systems is a necessity in protecting the common good.

## CHAPTER FIVE

# ADVOCACY AND EDUCATION

*"Ability to see the cultural value of wilderness boils down, in the last analysis, to a question of intellectual humility. The shallow-minded modern who has lost his rootage in the land assumes that he has already discovered what is important; it is such who prate of empires, political or economic, that will last a thousand years. It is only the scholar who appreciates that all history consists of successive excursions from a single starting-point, to which man returns again and again to organize yet another search for a durable scale of values. It is only the scholar that understands why the raw wilderness gives definition and meaning to the human enterprise."*

– Aldo Leopold
*A Sand County Almanac*

Education is intrinsically valuable for an individual, but it is also beneficial for society. Educated citizens are more likely to vote, to have better health habits, to get a job, and to be advocates. Today much emphasis is put on education as it relates to employment, but the original purpose of publicly funded education was not so narrowly conceived. In fact, the oldest public schools predate the industrial revolution. While there is an undeniable relationship between the numbers of educated citizens, the strength of the workforce, and the state of a nation's economy, an equally important objective for education is to prepare young people for citizenship. An inscription at the Boston Public Library states, "The commonwealth requires the education of the people as the safeguard of order and liberty." Here citizenship refers not to a person's legal status but rather to the relationship between a person and his or her community.

This distinction is important as it applies to education. Education creates citizens by revealing our shared principles—by allowing us to discover, cultivate, and express our values and exchange our ideas with others. Education helps us come up with solutions to the collective problems we face and levels the playing field among those with varying demographics.

There can be little advocacy without education. In order to want to protect something outside of ourselves, we first must value it, and in order to value it we have to learn about it. Certainly a lot of learning takes place during our formative years before any formal schooling begins, as it did for me on a small farm in rural Wisconsin. I distinctly remember how playing outside in the fields and woods around my parents' and grandparents' homes sparked my curiosity about and love of the outdoors. Formal schooling, however, enhances our understanding of both the human and nonhuman systems in which we live, work, and play, thereby instilling within us our values. While an individual doesn't automatically value everything he or she learns about, a person would have a hard time valuing something he or she knows nothing about.

In order to value wilderness, a person must learn many things about it, including the animals that live there, the ecological processes that take place there, its function in providing clean air and water, the natural evolution that occurs within its boundaries, and the impact of human activity within it. Young people might also learn to value wilderness when they are taught about the role it has played in our nation's history, the inspiration it has provided great authors and artists, or the therapeutic benefits it offers. Learning about the natural world and about humans' relationship to it is an essential component of learning to be a citizen. Maintaining the common good and finding common ground both require citizens to be critical thinkers, questioning deeply held beliefs and assumptions about the role wilderness plays in our society. Citizens must use their critical thinking skills to understand the positions of stakeholders in our collective action dilemmas.

Education prepares young people for citizenship by giving them the tools to contribute to both economic and social capital. A citizen has an obligation to work on behalf of his or her ideals in order to create a more perfect society; thus, a person who values wilderness should work to protect it. This advocacy can range from joining a wilderness society that aids federal agencies in delineating wilderness areas to voting for candidates who share their ideals, donating time or money to organizations that support their cause, or educating those around them about the importance of their convictions.

## Higher Education

Enrollment in institutions of higher education is increasing in the United States. As of 2013 just over 33 percent of young adults between the ages of twenty-five and twenty-nine had earned a bachelor's degree. (This varies, of course, by location and by socioeconomic status; in Massachusetts, for example, over 48 percent of this age cohort and almost 70 percent of those whose parents were in the highest income quartile had earned bachelor's degrees.) Universities that provide a liberal education expose students to a broad array of subjects that are important for a critical understanding of the world, but they also require young people to major in a field they are passionate about. The student of biology—or of physics, geology, political science, psychology, sociology, history, or many other fields—offers up fresh perspectives for wilderness preservation based on their own passions and areas of interest. Moreover, most public colleges and universities have a decidedly civic mission statement and recognize the importance of preparing young adults to be active and engaged members of the communities in which they live and work.

Creating environmentally educated citizens naturally leads to more citizens advocating on behalf of wilderness. One reason I love teaching young adults is their honest and unbridled intellectual curiosity. It is refreshing to talk with a student who just never thought about something like the relationship between humans and the natural world before. As a young adult it is easy to take for granted the current state of affairs, especially as it applies to wilderness. Young people today have not known a time without the Wilderness Act, the Clean Air and Clean Water Acts, or the Endangered Species Act. They don't remember the polluted Cuyahoga River starting on fire, or the near extinction of the bald eagle. The amount of designated wilderness areas has grown steadily over the course of their lifetime. Learning about environmental problems of the past and the policies put in place to solve them empowers budding wilderness advocates.

## Experiential Education

Formal education is essential in teaching children and young adults to care for the natural world. Yet it can be even more meaningful for those students to get out of the classroom and experience new ideas and viewpoints in person. Young people who get to hear directly from stakeholders about how a policy has affected them are much more powerfully affected than those who only listen to a lecture about it. I can't help but think of one of the ranchers my students had the chance to talk with about wolf reintroduction in Yellowstone. He made a deliberate decision to talk with them as they stood in the middle of his herd of cattle. Since we were there in the spring, many of the cattle were small—and very cute—calves. While a classroom discussion of cattle depredation might have sparked the interest of a few of those young adults, looking directly at the calves that were pastured in an area where wolf predation is possible brought home the point about the difficult relationship between cattle ranchers and wolf enthusiasts. Young people are more deeply affected when they are hear directly about a person's values in addition to the facts.

Exposure to the outdoors should be one of public education's fundamental goals. The more often young people can experience the natural world, the more deeply they will come to understand it and in turn value it. As a society we need to ensure that young adults, regardless of their socioeconomic status, race, or level of education, experience wild places through trips to our national parks or forests, state or county parks, local nature preserves, or other wild areas. We need these future advocates of parks and of wilderness.

## A Course on the Stewardship of Public Lands

I am one of a handful of individuals working on an exciting new national course about the stewardship of public lands. The course is referred to as "blended" because the content will be available online; instructors can assign and use the materials to inform their lessons in the classroom. This exciting project is an attempt to bring experiential education into the classroom, particularly in areas where young people do not have the opportunity to experience nature directly.

The team working on the course is gathering and creating documents, videos, position papers, literature, photographs, and stories from stakeholders in four public land controversies in Yellowstone National Park: brucellosis, wolf reintroduction, grizzly bear protection, and snowmobiling in the park during winter. All of these controversies directly or indirectly involve both wilderness and public lands. With this project we hope to expose young people to all sides of the controversies in Yellowstone and ask them to find common ground. No matter their viewpoint, young people should be able to find something they relate to and value. Seeking common ground will require the students to think like citizens of a place that is hundreds, even thousands, of miles away from them.

In an ideal world, our education system would be able to get every student of this new course out into the field locally to talk with stakeholders of environmental controversies. I encourage others to follow this model and create experiences and opportunities for young people to come into direct contact with stakeholders.

## Advocacy for Wilderness and for Education

Without concerned and committed citizens, wilderness would not exist. Therefore, wilderness is dependent on education, and supporting wilderness requires support for publicly funded education.

Since the Wilderness Act was passed fifty years ago, the dialogue about education in this country has shifted dramatically. Today education is often equated with getting a job. And while overall funding for education is shrinking, there is a little more money in the science, technology, engineering, and mathematics (STEM) fields of study. To be sure, these are important fields, and young people who pursue careers in them will undoubtedly make important contributions to our society. At the same time, however, there has been a devaluation of the humanities and social sciences in recent years. Yet if citizenship is important for democracy, these fields are imperative. As an instructor of political science courses I have been truly shocked by how few young people know even the basics about how government works. Civics education is vital to wilderness protection. But so too are English, art, sociology, and history. The way young people develop values is directly related to how they perceive the world through different sets of spectacles.

Certainly every young adult needs to learn about biology, chemistry, geology, and physics in order to better understand the world he or she lives in. Every young adult also needs to learn about the relationship between humans and the land. This requires studying literature, history, and anthropology. To protect the natural world, citizens of all ages must learn to work together, and this necessitates learning about psychology and sociology. They also must learn about how and why individuals protect the common good and how to create laws and institutions. This requires education in philosophy and political science. They must be able to go out and share their ideas through speeches, writings, and performances, requiring the study of communication, art, English, and theater. Students of all ages should be exposed to literature about the environment and be encouraged to explore how that literature relates to their own experiences. I can only imagine what might happen if every student was required to read *A Sand County Almanac* and if every student was able to think critically about the relationship between humans and the land. The need for this type of broad, humanities-based education is urgent.

It is especially important for young adults to take seriously the cause of protecting wilderness and working with fellow citizens to find common ground. According to Jim Kurth, chief of the National Wildlife Refuge System, in twenty to thirty years we will likely designate our last wilderness area. The issue for the young generation, according to Kurth, is how we as a country prioritize wilderness after that point.

The world population has more than doubled since the Wilderness Act was passed in 1964, putting pressure on wilderness areas. Even in the United States, a country with relatively slow population growth, our lifestyle demands tremendous energy use and in turn resource extraction. The American lifestyle is also dependent on a healthy economy. In the most recent economic downturn we saw state government easing environmental regulations to help jumpstart the economy and reduce unemployment in local communities. Wisconsin, for example, eased environmental regulations for a proposed $1.5 billion open-pit iron ore mine in the northern part of the state, near the Chequamegon National Forest and only six miles from Lake Superior. Proponents of the mine appeal to the potential for economic growth; opponents cite fears that the mine threatens the environment and public health. We need to educate young adults so they will seriously consider the consequences of such actions and to ensure that they become advocates on behalf of all things wild.

## Vignette
# Young People in Yellowstone

In the summer of 2010 I took eight young people to Yellowstone National Park. All were political science or public administration majors at the University of Wisconsin–La Crosse, where I teach, and all had signed up for a special course through the Yellowstone Association (YA) that focused on the stewardship of public lands. I had taken a faculty version of the course a year earlier. The course introduced young people to public land controversies in the Greater Yellowstone Ecosystem (GYE), introduced them to stakeholders from all sides of these controversies, and challenged them to find common ground among diverse interests.

There were more thought-provoking moments during this six-day adventure than I can recount. One involves the transformation in one student's attitude, brought about by exposure to ideas that challenged his worldview. This student was a bit of a challenge in the classroom; he thoroughly enjoyed playing devil's advocate for all to hear, he tended to call things that challenged his worldview "stupid," and he liked being the center of attention. While I have a playful attitude in class and don't mind some banter, I was concerned he would disrespect the stakeholders we were going to meet with—particularly those with whom he didn't agree. These people spoke with us voluntarily, and I didn't want one student's attitude to discourage them from talking with other groups.

Not only did the student show proper respect to the speakers, afterward he acknowledged some of the good points made on opposing sides of the issues. I could hardly believe my ears when I heard him express support for an advocate from an organization that works to protect bison—not an organization I would have expected him to find any agreement with. This experience confirmed for me that having young people talk to those directly affected by policy can change their attitudes and open their minds, paving the way to common ground.

In another eye-opening experience, our group took a day hike up to pens where wolves had been kept before they were reintroduced to Yellowstone a decade earlier. During the hike our YA educator recounted the reasons why wolf reintroduction has been so controversial. While park scientists viewed wolves as a natural part of the ecosystem that could help restore ecological balance, some in the community worried about the wolves' effect on Yellowstone's elk population and on herds of cattle just outside the park. The educator told us that the pens had been protected by armed guards so anti-wolf advocates would not harm the animals or the enclosures. This very bright and chatty group of young people talked and asked questions all the way there. When we reached the pens, however, everyone became quiet. I watched the students' faces as they pondered the enormous undertaking wolf reintroduction had been. They were beginning to realize how an idea can turn into a policy implementation quagmire. Later in the week, as the students spoke with a local rancher about his losses due to wolves, I could tell the experiences of the week were weighing heavily on their minds. These young people had been able to see policy in action, and now they were keenly aware of its consequences.

Although the students didn't have time to go into a designated wilderness area, they did get to see a once-in-a-lifetime wildlife scene unfold in front of them. One afternoon we were busy watching some cute grizzly cubs play in the distance (via spotting scopes, of course). The cubs were rolling around and biting at each other as their mother lay nearby. Suddenly, a herd of bison decided to cross the road where our bear jam was located. We had no sooner loaded all of the young people onto the bus for their safety when the YA instructor starting imploring us to get them back off the bus. While the humans were distracted by the bison, the mother grizzly had traversed the landscape toward a mother elk and her calf. The mother elk called and stomped her feet, but it was not enough to keep the grizzly away from her calf. The young people watched the grizzly take down and eat the calf (I, of course, looked away). They had probably seen this on nature shows before. But I doubt any of them had or ever again see something like this in person. I know the young people learned something about valuing the natural world on that day. Nature isn't always pretty, but there is value in knowing that wildness exists. In summarizing the trip, one of my students, Tyler Burkart, wrote:

> Professor Arney encouraged me to explore this interest further by taking an educational trip to Yellowstone Park. Overall, I never considered myself an outdoorsman. Despite this fact, I believe going to Yellowstone and studying public policy there was one of my most positive experiences in my academic career. It was extremely beneficial to have the firsthand experience and chat with the Yellowstone locals about issues that impact them on a daily basis, such as brucellosis, the wolf population, the consequences of having non-native fish in Yellowstone, and efforts to reduce pollution in the area. It was an added bonus to have the remarkable scenes around you every day at Yellowstone Park. Studying these challenging issues really opened my eyes

to other public policy issues that I may not face in the Midwest region, such as water usage in the Southwestern region and natural disasters in the Gulf of Mexico. Furthermore, this experience has inspired me to take more challenges and have a greater appreciation for Mother Nature.

# CHAPTER SIX
# WILDERNESS AND OUR NATIONAL PARKS AND FORESTS

*"An administrator of public lands containing remnants of wilderness should be aware of the fact that the richest values of wilderness lie not in the days of Daniel Boone, nor even in the present, but rather in the future. The administrator has a double responsibility; to keep some wilderness in existence, and to cultivate its qualitative enjoyment."*

– Aldo Leopold

*Wilderness as a Form of Land Use*

During the fight for congressional designation of wilderness between 1956 and 1964, both National Park Service and US Forest Service officials argued against the Wilderness Act because it limited administrative discretion about where the boundaries of wilderness would be located. Some authors have suggested that these officials were more concerned about their freedom to change boundaries than they were about protecting wilderness. I would suggest that these types of power shifts are always rocky. Most people who follow a career path into the park or forest service do so because of their love of nature and not for a career that promises lots of political power. I have been only a tourist within park and national forest boundaries, but from my perspective the National Park Service and US Forest Service are great stewards of both wild areas and wilderness.

## Wildness for the Masses

I remember thinking an environmental philosophy professor was something of an elitist when he said public parks are good for the masses because they keep crowds away from the more remote areas, leaving them for the "real" outdoors enthusiasts to enjoy. He may have been joking or being sarcastic to make a point. But his words made me think about the place of humans in wild areas. Too many human visitors can damage an area and thereby make it less wild. However, if no humans have access to a given area, arguments for its protection will fall on many deaf ears.

The truth is that wilderness protection doesn't have to be an all-or-nothing proposition. National parks and forests offer citizens the opportunity to see something relatively uncultivated by humans. I have been to only about a dozen of the fifty-nine national parks and have stopped in only a few dozen of our national forests (though I have driven through many more of them), yet in all cases they offered landscapes and wildlife that I could not have experienced except within their boundaries. National parks and forests are good for the masses, not because they keep people away from wild areas, but because most citizens are happy enough with simply experiencing the outdoors. There aren't that many folks who want to strap on a backpack and disappear into the wilderness for a few days.

National parks and forests give citizens the opportunity to experience something wild while still allowing them to feel safe and comfortable.

These "wild" (but not wilderness) areas offer more than just a photo opportunity for citizens. They provide a place for creating and refining our values. Park visitors might be captivated by the scenic vistas on Trail Ridge Road at Rocky Mountain National Park, or the lower falls in Yellowstone National Park, or the rock formations in Arches National Park. They have the opportunity to appreciate the history and marvel at the wildlife, the scenery, and the sheer uniqueness of many of the landscapes and features found within park borders. Who can forget a hike down to the cliff dwellings at Mesa Verde or the length of the drive around Crater Lake? Given that nearly 275 million people visit national parks each year, it is clear that citizens have fallen in love with these areas. As they do, they begin not only to value the places themselves but also to deepen their values of the wildness found within them. A person must value wildness before he or she can value wilderness. The national parks and forests are imperative in the movement to save wilderness because they are citizens' windows into the wild. Parks and forests give us all the opportunity to play in the wild—to hunt, to fish, to hike, to draw, to photograph, and to imagine.

National parks and forests also serve as gateways to wilderness. Seventy-seven percent of wilderness areas are found within their borders. Those who do venture into the wilderness often use the national parks or forests as an entry point.

## Education within Park and Forest Borders

National parks and forests are also stewards of wilderness because they educate the public about wilderness. They offer publications, interpretive guides, classes, and other instruction about the flora, fauna, and landscapes within their borders—half of which happens to be designated wilderness. Rangers teach people how to treat the land and how to behave in their interactions with wildlife. They relay the history of the area and describe the forces shaping the parks today. This education provided by the park and forest service officials propagates values.

## Yellowstone

As you have guessed by now, Yellowstone has a special place in my heart. For me going to Yellowstone is like going home. Most of my trips there have involved the Yellowstone Association, Yellowstone's official nonprofit education partner. After I started going there regularly, a friend told me she would never go to Yellowstone again because park officials had wrecked the Yellowstone of her youth. The Yellowstone of today, she said, is filled with boardwalks and paved paths and too many tourist concessionaires. While I could understand the sentiment, the Yellowstone that I know and love is so much more than these things. The majority of Yellowstone visitors stay on the figure-eight road and stop only at featured areas, such as Old Faithful and Artist Point. They are getting what why want out of the park, and even if their visit is short, they are learning to value the park and all it contains, even though they're not seeing its most extraordinary features.

What I love about Yellowstone are the public controversies and policy issues that arise within its boundaries and include everything from wolf recovery to brucellosis in bison. I enjoy seeing these animals in their native habitat but also talking to the people invested in the policy quagmires. I love learning about these things through the eyes of my students. I convinced my friend to come along when I took a group of young people, and, sure enough, she fell back in love with Yellowstone. They key, we agreed, was working with a trained educator from the Yellowstone Association. Wolf watching is great, but wolf watching with an educator who can tell you the history and struggles of the particular pack you are watching leads to a deeper understanding and appreciation. Every citizen who has the opportunity should take a class with a park ranger or a trained Yellowstone Association educator. The park's features are spectacular, but putting them in context increases the quality of the experience and increases visitors' understanding and enjoyment.

## National Park and Forest Service Mission

It would be shortsighted to argue that our national parks and forests must be defenders of wilderness at all costs. While wilderness protection is part of their mission, it is clearly not all of it. National parks exist so that citizens can witness and enjoy nature, and park officials offer a very important instructive service to citizens. National forests educate citizens about how to use resources in a sustainable manner. These activities may not be congruent with being a premier wilderness protection agency, but they are important for the common good. As these agencies educate citizens, they instill values—the same values that are needed to protect wilderness. While these agencies may have opposed the Wilderness Act in the past, today both the National Park Service and the US Forest Service are essential stewards of and gateways to wilderness.

### Vignette

## Rollin' on a River (or, Sunglasses at Night)

Every time I hear the song "Sunglasses at Night" by Corey Hart, I am taken back to a river in rural Montana. In 2006 I was a graduate school research assistant collecting and analyzing data for the Indian Housing Block Grant, part of the Native American Housing Assistance and Self Determination Act of 1996. That summer a colleague and I traveled to a Tribal Housing Authority (THA) office in Montana to collect information and set up a system to make future data collection easier from a distance. Our hosts were terrific, and with everyone's hard work and dedication we finished the project early. A local resident affiliated with the THA asked my colleague and me if we would like to spend the afternoon on a boat ride down a river. Would we? That would be terrific!

We changed into shorts, T-shirts, and sunglasses and headed out to meet our host, stopping at a gas station to grab some chips and a candy bar to tide us over until dinner since we had worked through lunch. We drove on to what would be described most accurately as the middle of nowhere to meet our host and a ranger who would drive us to the boat landing. As we hopped into their vehicle, I noticed a few signs that something might go wrong on this adventure—signs I chose to ignore. First, our host was still wearing a suit, not exactly the attire an experienced river guide wears. Then, when our host and the ranger unloaded the boat we'd be using, we saw that it was a rowboat with a bouncy bottom, not the ideal vessel for river navigation. The third red flag came when our host proclaimed, "I looked for some young gentlemen to row you ladies, but no one was available, so one of you will have to row." Luckily my colleague had some rowing experience.

As we started down the river, any thoughts of trouble melted away. Except for the fact that our host was now helping himself to our chips (our only sustenance for the day), the trip down the river was great. The current was slow, and the scenery was amazing. We were also all engaged in a very charming conversation about everything from spirituality to treaty and water rights to our personal histories. After what seemed like about an hour, we saw the ranger who had dropped us off waving us over to shore. We had reached our takeout point. Our host asked, however, if we would like to keep going. "How far have we gone?" we asked. "About three miles," he answered. "How far is the next takeout?" "About another three miles," he replied. Sure, we would love to keep going! What we didn't know—and could not have known—is that we had gone only one mile, and the next takeout was more than six miles downstream. Considering that we had gone only a mile in an hour and it was already about five PM, you can imagine what happened.

I remember feeling uneasy as we watched our ranger waving from shore. But we trusted that our host was the expert. The conversation was still lovely, but as we floated on I began looking at the shoreline and wondering when and where we would be able to get out. The current remained slow, and some parts of the river were so shallow that we had to get out to push the boat. Our host started saying things like "I haven't floated this in years. I don't remember this part of the river!" and "The river sure has changed a lot." As the sun started setting it began to get cold, and getting out of the boat to give it a push became more than a nuisance. We were freezing. All we had to keep us warm were our life jackets, which were getting wet. I vividly recall seeing the lights of a house in the distance and wondering if we should call for help. Yet our host seemed calm, so we stayed on the water.

It wasn't long after that that we entered a small canyon of sorts. The rock walls along the river weren't very high, but now there was no way we would be able to get up on shore. It got dark, really dark, and fast. My colleague was still rowing. She was having a hard time seeing, but her sunglasses were also her prescription glasses, so she couldn't take them off—without them she would not have seen anything at all. We could hear the sound of rushing water but no longer felt confident that it was simply shallow water over rocks. Every time we had to get out to push the boat, it was a risk, as we couldn't see what we were stepping into. My friend inquired about the possibility of hiking out. But we didn't know where we were, we couldn't see anything, and hiking could be just as dangerous as staying on the water. Of course, I also imagined all of the bears out there waiting to eat us. Our situation got really scary when we began to see thunder and lightning off in the distance. Our host asked us to pull to shore so he could relieve himself. While he was out of earshot, my colleague and I wondered aloud whether it would be smartest to pull the boat out of the water and try to use it as shelter from the weather.

It was then we saw a spotlight on the rock outcrop up ahead. Hurry, hurry! We were paddling toward the light when it disappeared. While we were certain someone was looking for us, we knew we had missed them. We continued slowly downstream. A little while later the lights reappeared, and then we heard, "I see them!"

It was almost midnight by the time we got back to where we were staying. We were hungry, but sleeping was more important than eating. First I phoned Jeremy to tell him we were okay. Of course, he was back at home in Colorado and had had no idea we were even missing.

I hadn't fully comprehended how much trouble we were in until after we had been rescued. Although

I'd had other brushes with danger while adventuring outdoors, this was the only time I contemplated my mortality in the face of being lost in the wild. And yet, here I sit, writing this book, lucky to have come out of that experience unharmed, even though I made some pretty dumb mistakes: I didn't listen to my instincts, I didn't pack the right gear, I was tired and weak from not being properly nourished, I trusted someone I barely knew to be my guide.

As humans, we make mistakes. We don't always treat the land as we should. In the course of our nation's history we have made shortsighted management decisions. And yet wilderness continues to exist. Wild areas, both those with and those without human footprints, exist in this country, and both kinds deserve protection. I am admittedly a conservationist, and I want to see wilderness set aside for both biological and intrinsic reasons. Those areas with a human footprint, however, should not be overlooked or written off as not valuable and worth protecting. Accessible wild areas are necessary if we want to maintain wilderness.

The opponents of the concept of wilderness raise a valid point when they call it idealistic. In some ways it is. The ideal of wilderness can result in the devaluation of areas that fall outside of the political definition. Not every wild area in this country is going to be preserved free of human impact. It is simply not possible, and given the many competing needs and values in our society, it is not desirable. It is only by acknowledging this reality that we can move forward, compromise, and find common ground. It is how we will rescue wilderness.

# Conclusion
# The Future of Wilderness

*"All ethics so far evolved rest upon a single premise: that the individual is a member of a community of inter-dependent parts. His instincts prompt him to compete for a place in that community, but his ethics prompt him also to co-operate (perhaps in order that there may be a place to compete for). The land ethics simply enlarges the boundaries of the community to include soils, waters, plants, and animals, or collectively: the land."*

– Aldo Leopold
*A Sand County Almanac*

The future of wilderness is in good hands. Wilderness advocates have done a tremendous job of institutionalizing the concept of wilderness in the United States and have been successful in persuading members of Congress to declare 757 wilderness areas comprising over 109 million acres. The United States is one of only three countries whose government designates areas as wilderness and protects them as such, and wilderness areas are now spread across the country. Forty-four states and Puerto Rico have designated wilderness areas. In fact, nearly 16 percent of Alaska and 15 percent of California are declared wilderness areas. These states are recognized for their biodiversity, and the wilderness areas within their borders are of value to all Americans.

Citizens have also increasingly become advocates for all things wild. There are more wilderness advocacy organizations today than ever before. The website *Wilderness.org* includes a list of more than 370 organizations that work to support wilderness. Moreover, organizations such as The WILD Foundation work to protect wilderness around the world.

## The Need for Wilderness Today

The world is drastically different than it was in 1964. The Internet, mobile devices, and social networking have moved our society in new directions. Not only have these innovations changed our relationship with the world, they have changed the culture of education. I occasionally run a contest in the classroom in which I allow young people to use their phones to look up an answer. I ask a random question, such as the name of the twelfth vice president of the United States or the state bird of Kentucky. My students can usually tell me the answer in seconds. With information available instantly at our fingertips, those of us in education are no longer the suppliers of facts. We reveal facts to young people as part of instruction, to be sure, but since young people can look up almost anything on a mobile device in a matter of seconds, we can better use our classroom time to teach young people to be critical thinkers and disseminators of information. Experiential education is an imperative part of that process. If we want young people to become citizens who advocate on behalf of the environment and wilderness, then we must help them experience it directly.

This is exactly why having wilderness is important for our future. We need wilderness to exist in order to experience it. It is mainly through exposure that we cultivate values for wild things—values that in turn motivate us to adopt and implement policy. We simply can't inspire the building of values by looking at a picture on a mobile device. While the background on my laptop may be photo of Yellowstone, I know that I am not experiencing Yellowstone when I look at it. Someone else might admire the photo, but without experiencing it directly that person is unlikely to come to value it.

The downside of all this connectivity, of course, is the potential for becoming overly connected, at the cost of experiencing nature. While it is really nifty to be able to post a picture taken at Artist Point on Facebook or Instagram, for example, those who see it can feel less urgency to experience it for themselves. I once traveled with a group of high schoolers to do service work near the Pine Ridge Reservation. As part of the journey we spent a night in Rapid City, South Dakota, and naturally I planned an excursion to see Mount Rushmore. The majority of the young people, however, opted to go mini golfing instead—because, they explained, they had seen Mount Rushmore in pictures.

This is the challenge for educators and wilderness advocates alike: to motivate citizens to experience and value wild places. In doing so, we should encourage others to not just visit the wild areas but also to talk to their fellow citizens whose lives and livelihoods are dependent on those areas. As citizens we should all be having conversations with the stakeholders of wilderness. When thinking about wolf reintroduction in Yellowstone, for example, the stakeholders include conservationists, ranchers, park scientists, and dozens of local businesses that could earn millions of dollars from wolf watchers who visit the park. Different stakeholder groups are likely to have different viewpoints. Finding common ground will require respecting all of them.

### Wilderness and Citizenship

Given my defenses of local economies and the agricultural and rural lifestyle, one might think that I more or less subscribe to a sustainable development philosophy. The truth is that I am a wilderness advocate to the core. I believe wilderness is intrinsically valuable and ought to be saved for no other reason than it exists. I am also an advocate for our common good, however. There is a reason that this book is titled *Wilderness and the Common Good* as opposed to *Wilderness as the Common Good.* Wilderness is part of our common good, and dedicated citizens should fight to protect it.

As Americans we should be proud of a political system that was able to produce a law as visionary as the Wilderness Act. But this complex political system also depends on citizens who have and are willing to share different viewpoints. We are certainly not a society of one mind. I like to remind young people that what makes our society great is the freedom to have and express a variety of opinions. As I devote my attention mainly to defending the environment, I am glad other individuals' priorities include education, health care, and national defense. I don't put much thought, time, or energy into these policy areas, but I benefit from their existence. We all do. It is to the great benefit of our common good that we have a variety of interests, because putting a lot of energy into one area often comes at the expense of others.

I see the relationship between wilderness and citizenship as being symbiotic. Citizens need wilderness just as wilderness needs citizens. Wilderness is a common good on which we all depend, much like clean air and water. The historical roots of our country remind us about the role wilderness has played in the creation of our national identity. We need wilderness because we need a place for the generation of raw ideas along with raw materials. The good news is that the Wilderness Act made an excellent start in protecting wildlands for our future. But there is work to be done. Luckily we can all practice citizenship skills in an effort to protect wilderness for the future.

## Moving Forward

On the path to wilderness protection, we need to fight the urge to demonize those who have viewpoints different from our own. I am especially bothered by environmentalists and wilderness advocates who try to paint private citizens, such as ranchers and families whose income is dependent on logging or mining, as the enemy. The truth is that these people often value the land just as much as their counterparts do. Opponents of wilderness designations often point to a variety of other activities that they believe would enhance the area for common use—some of them individual pursuits, such as mountain biking, and some of them collective, such as a development that would lead to job growth. These individuals also value the common good. But while I am leery about sustainable development in wilderness areas, I understand that we can't designate every wild area as wilderness. There must be areas reserved for these other functions of society. We are all dependent on the economic growth that accompanies many extractive industries. If as environmentalists we don't support these industries, then we need to look for other ways to respond to these societal needs, rather than simply making enemies of our fellow citizens. One example is to consider our own use of resources and how it affects protecting wild areas and to make changes to reduce our impact on the environment. Another is to support the creation of green jobs. The pursuit of green jobs requires continued innovation and education. Once again, the relationship between protecting wild areas and education are inextricably linked.

When our opinions about the function of a wild area differ, finding a shared value represents an ideal situation. Moving ahead with shared values demands respect, but it also requires compromise. Too many elected officials and citizens alike seem to see compromise as a loss. Compromise is never a loss if both sides gain something. Of course, compromise also compels both sides to give up something. In political science we use the word *logrolling* to describe a situation in which one faction gets its way in policy issue A in exchange for the other faction getting its way in policy issue B. There is nothing inherently unethical about logrolling. It happens all the time in everyday life. As members of an interlinked community, there is no reason not to push for this strategy. If a mine must go in area A, then area B should be set aside as wilderness or at the very least as a local or state park where it will be afforded some protection.

Wilderness protection requires citizens who both understand and respect the opinions of others. As Aldo Leopold so aptly notes, we are all "members of a community of interdependent parts." We can and must work together to solve the collective action problems we face as a society. Those in the wilderness community must learn to work with, and not against, those who favor sustainable development. These groups already have shared values. Sustainable developers, by definition, want to use the land in a way that does not ruin it for our future. Sustainable developers and environmentalists have more in common than either does with individuals who don't care about the future of the environment. It is by working for the common good that we can further the legacy of the Wilderness Act.

## References

ABC News/Planet Green/Stanford Poll, July 2008. Retrieved Jun-28-2013 from the iPOLL Databank, The Roper Center for Public Opinion Research, University of Connecticut. http://libweb.uwlax.edu:5641/data_access/ipoll/ipoll.html

Callicott, J. Baird, and Michael P. Nelson, editors. *The Great New Wilderness Debate*. Georgia: The University of Georgia Press, 1998. Print.

Chief Luther Standing Bear. "Indian Wisdom." In *The Great New Wilderness Debate*, J. Baird Callicott and Michael P. Nelson, editors. Georgia: The University of Georgia Press, 1998. 154–98. Print.

Climate Change and Global Poverty Survey, March 2009. Retrieved Jun-28-2013 from the iPOLL Databank, The Roper Center for Public Opinion Research, University of Connecticut. http://libweb.uwlax.edu:5641/data_access/ipoll/ipoll.html

CNN/Opinion Research Corporation Poll, July 2008. Retrieved Jun-28-2013 from the iPOLL Databank, The Roper Center for Public Opinion Research, University of Connecticut. http://libweb.uwlax.edu:5641/data_access/ipoll/ipoll.html

Democracy Corps Poll, June 2008. Retrieved Jun-28-2013 from the iPOLL Databank, The Roper Center for Public Opinion Research, University of Connecticut. http://libweb.uwlax.edu:5641/data_access/ipoll/ipoll.html

Gallup Poll, May 2008. Retrieved Jun-28-2013 from the iPOLL Databank, The Roper Center for Public Opinion Research, University of Connecticut. http://libweb.uwlax.edu:5641/data_access/ipoll/ipoll.html

Leopold, Aldo. "The Last Stand of the Wilderness." *American Forests and Forest Life* 31.382 (1925): 599–604. Print.

Leopold, Aldo. "Wilderness as a Form of Land Use." *The Journal of Land & Public Utility Economics* 1.4 (1925): 398–404. Print.

Leopold, Aldo. *A Sand County Almanac*. New York: Ballantine, 1949. Print.

Leopold, Aldo. *Round River: From the Journals of Aldo Leopold*. Luna B. Leopold, editor. New York: Oxford University Press, 1953. Print.

Leopold, Aldo. *The River of the Mother of God and Other Essays by Aldo Leopold*. Susan L. Flader and J. Baird Callicott, editors. Madison: University of Wisconsin Press, 1991. Print.

Leopold, Aldo. *For the Health of the Land: Previously Unpublished Essays and Other Writings*. J. Baird Callicott and Eric T. Freyfogle, editors. Washington, DC: Island Press, 1999. Print.

Muir, John. *The Wilderness World of John Muir*. Edwin W. Teale, editor. Boston: Houghton Mifflin, 1954. Print.

Nelson, Michael P. "An Amalgamation of Wilderness Preservation Arguments." In *The Great New Wilderness Debate*, J. Baird Callicott and Michael P. Nelson, editors. Georgia: The University of Georgia Press, 1998. 154–98. Print.

Rolston, Holmes. "Feeding People versus Saving Nature?" In *World Hunger and Morality*, William Aiken and Hugh LaFollette, editors. Englewood Cliffs, New Jersey: Prentice Hall, 1996. 248–67. Print.

Scott, Doug. *The Enduring Wilderness*. Golden, Colorado: Fulcrum Publishing, 2004. Print.

Unified Arctic Campaign Survey, June 2008. Retrieved Jun-28-2013 from the iPOLL Databank, The Roper Center for Public Opinion Research, University of Connecticut. http://libweb.uwlax.edu:5641/data_access/ipoll/ipoll.html

## About the Author

**Jo Arney** is an associate professor of political science and public administration at the University of Wisconsin–LaCrosse. Her main area of teaching and research interest is in environmental politics and policy. Jo is also one of the lead scholars in the creating of a nationwide blended course about the stewardship of public lands being developed by the American Association of State Colleges and Universities.